Milk, Modernity and the Making of the Human

This book undertakes a critique of the pervasive notion that human beings are separate from and elevated above the nonhuman world and explores its role in the constitution of modernity.

The book presents a socio-material analysis of the British milk industry in the late nineteenth and early twentieth centuries. It traces the dramatic development of the milk trade from a cottage industry into a modernized and integrated system of production and distribution, examining the social, economic and political factors underpinning this transformation, and also highlighting the important roles played by various nonhumans, such as microbes, refrigeration technologies, diseases, and even cows themselves. Milk as a substance posed deep social and material problems for modernity, being hard to transport and keep fresh as well as a highly fertile environment for the growth of bacteria and the transmission of diseases such as tuberculosis from cows to humans. *Milk, Modernity and the Making of the Human* demonstrates how the resulting insecurities and dilemmas posed a threat to the nature/culture divide as milk consumption grew along with urbanization, and had therefore to be managed by emergent forms of scientific and sanitary knowledge and expertise.

Milk, Modernity and the Making of the Human is an ideal volume for any researcher interested in the hybrid socio-material, economic and political factors underpinning the transformation of the milk industry.

Richie Nimmo is a Lecturer in Sociology at the University of Manchester, UK. His research explores the ambiguous status of nonhumans in modern knowledge-practices and the constitution of 'the social' across materially heterogeneous relations, systems and flows.

Culture, Economy and the Social

A new series from CRESC – the ESRC Centre for Research on Socio-cultural Change

The *Culture, Economy and the Social* series is committed to innovative contemporary, comparative and historical work on the relations between social, cultural and economic change. It publishes empirically-based research that is theoretically informed, that critically examines the ways in which social, cultural and economic change is framed and made visible, and that is attentive to perspectives that tend to be ignored or side-lined by grand theorising or epochal accounts of social change. The series addresses the diverse manifestations of contemporary capitalism, and considers the various ways in which the 'social', 'the cultural' and 'the economic' are apprehended as tangible sites of value and practice. It is explicitly comparative, publishing books that work across disciplinary perspectives, cross-culturally, or across different historical periods.

The series is actively engaged in the analysis of the different theoretical traditions that have contributed to the development of the 'cultural turn' with a view to clarifying where these approaches converge and where they diverge on a particular issue. It is equally concerned to explore the new critical agendas emerging from current critiques of the cultural turn: those associated with the descriptive turn for example. Our commitment to interdisciplinarity thus aims at enriching theoretical and methodological discussion, building awareness of the common ground that has emerged in the past decade, and thinking through what is at stake in those approaches that resist integration to a common analytical model.

Series titles include:

The Media and Social Theory (2008)
Edited by David Hesmondhalgh and Jason Toynbee

Culture Class Distinction (2009)
Tony Bennett, Mike Savage, Elizabeth Bortolaia Silva, Alan Warde, Modesto Gayo-Cal and David Wright

Material Powers (2010)
Edited by Tony Bennett and Patrick Joyce

The Social after Gabriel Tarde: Debates and Assessments (2010)
Edited by Matei Candea

Milk, Modernity and the Making of the Human (2010)
Richie Nimmo

Cultural Analysis and Bourdieu's Legacy (forthcoming)
Edited by Elizabeth Silva and Alan Ward

Creative Labour – Media Work in Three Cultural Industries (forthcoming)
Edited by David Hesmondhalgh and Sarah Baker

Rio de Janeiro: Urban Life through the Eyes of the City (forthcoming)
Beatriz Jaguaribe
Sharon Zukin, Brooklyn College City University New York/ Graduate School, City University of New York

E·S·R·C ECONOMIC & SOCIAL RESEARCH COUNCIL

CRESC Centre for Research on Socio-Cultural Change

Milk, Modernity and the Making of the Human

Purifying the social

Richie Nimmo

Routledge
Taylor & Francis Group
LONDON AND NEW YORK

First published 2010
by Routledge
2 Park Square, Milton Park, Abingdon, Oxon OX14 4RN

Simultaneously published in the USA and Canada
by Routledge
270 Madison Avenue, New York, NY 10016

Routledge is an imprint of the Taylor & Francis Group, an informa business

Typeset in Times by
RefineCatch Limited, Bungay, Suffolk

British Library Cataloguing in Publication Data
A catalogue record for this book is available from the British Library

Library of Congress Cataloging-in-Publication Data
Milk, modernity and the making of the human : purifying the social / edited by Richie Nimmo.
p. cm.
Includes bibliographical references.
ISBN 978-0-415-55874-7 (hbk.)–ISBN 978-0-203-86733-4 (ebook)
1. Milk trade–Social aspects–Great Britain. 2. Dairying–Social aspects–Great Britain. 3. Human-animal relationships.
4. Humanism. I. Nimmo, Richie.
HD9282.G72M55 2009
306.3–dc22 2009034688

ISBN10: 0–415–55874–3
ISBN10: 0–203–86733–5

ISBN13: 978–0–415–55874–7 (hbk)
ISBN13: 978–0–203–86733–4 (ebk)

Contents

Preface

We are living in a time of unprecedented crisis. It is true that radical or critical social scientists have always tended to identify their own time as a time of major crises of one sort or another. But this time I think is different. The end of modernity, which has for so long been prematurely hailed, celebrated and discussed in abstract discourse, is at last upon us materially in the form of a world crisis, not merely of capitalism, nor of society, but of nature. Far from some cultural shift towards a postmodern condition, the conditions of existence of modernity as we have known it are quite simply disappearing, and rapidly. Unstoppable climate change and ecological exhaustion are the material culmination of the contradictions of a global modernity, its economics, its science, and the nature it has produced. This seems to me manifestly the defining problem of our time, but one which is not yet really understood in its fullest implications. 'Environmentalism' is far too narrow a response; so are most forms of green politics and certainly green consumerism. It should by now at least be clear that no such re-arranging of the deckchairs will do. Catastrophic climate change is already inevitable, and rather than making a new cultural class politics out of doing far too little far too late, we should focus instead upon two things: how to mitigate its impact, and how we came to be in this situation.

The first question is essentially pragmatic, although nonetheless staggering in the enormity of what is required, demanding social and material re-organization of a scale, speed and degree of global co-ordination unlike anything previously attempted; to consider that it may well prove beyond the capacity of our species to accomplish is not mere pessimism. The second question is even more revolutionary in its implications. Properly understood, it is much more than just a matter of fossil fuels and carbon emissions, but reaches into how a particular conception of humanity in relation to non-humanity – that is, how a specific version of 'the social' vis-à-vis 'nature' – has been inscribed and entrenched by modernity as an epoch-defining form of life. At the deepest level, this entails nothing less than a fundamental re-thinking of what it means to be human. We find ourselves in a conjuncture in which it is increasingly unsustainable to understand human beings in terms of any sort of culture–nature dualism. Historical conditions have precipitated a crisis of humanist ontology through the proliferation of recalcitrant natures capable of subverting modern forms of socio-natural order by asserting their own agency. One thinks not only of climate change itself but also of its many 'feedback' effects, of associated new vectors of zoonotic disease, and of accelerating antibiotic resistance, as ominous signs of the return of a repressed 'nature'. Similarly, such developments as the growth of social movements around various 'politics of nature',[1] the emergence of significant public concern for animal welfare quality in food production networks, as well as the implications of developments in the life sciences, all in their different ways contribute to a historical situation in which the question of how we are to live, how we are to define and practise our humanity in a world of nonhuman others; in short, how we are to *be* human, is contestable now as never before. It is to the transformational possibilities, however remote, which are signalled by that existential politics, made thinkable by the crises of our time, rather than to any more narrowly scholastic, environmentalist, or animal-rights agenda, that this book aims ultimately to contribute.

Though its underlying politics then are broad and overarching, the book's empirical subject-matter is quite specific. It is a historical analysis of the modern British dairy industry, and the liquid milk trade in particular, which I trace in detail from the earliest days of its transformation into a modern industry in the 1860s, through to the dramatic commercializations of the 1920s and 1930s. I decided to focus upon the dairy industry above all because I think that milk itself is fascinating. Unlike meat, it has not previously been subjected to any sustained analysis in terms of the heterogeneous relations that it embodies and mediates.[2] As a commodity which is not simply an animal's body – and therefore an animal that has been quite literally 'objectified' – but is instead the organic product of a still *living* animal, milk is highly suggestive. It offers a unique way to examine the role of hybrid human–nonhuman relations in underpinning the modern commodity form, as well as the means by which modernity struggles to efface, contain, or 'purify' these trangressive hybridities and the corporeal testimony of animals and other nonhumans which they so strikingly articulate. Whereas meat has for countless centuries been cooked before being eaten, providing an effective means of both material and symbolic purification, and ensuring the transformation of its animal 'nature' into an object of 'culture' before its consumption by humans, the equivalent purification of milk is entirely a product of the late nineteenth and early twentieth centuries. Indeed, prior to the development of refrigeration technology, microbiology, systematic milk testing and pasteurization, milk was an ideal medium for bacterial growth and a potent vehicle for the transmission of various diseases, both between humans and between cows and humans. Milk has therefore been deeply and materially problematic for modernity in ways that meat has not.

Certainly meat can and has been critically problematized, theoretically, ethically, and in all sorts of ways, as the dramatic growth of vegetarianism illustrates; but the incorporation of meat consumption into the modern order has not demanded anything like the extent of technical, material and ontological restructuring and negotiation as has the normalization of milk consumption. This can in part be attributed to the historical difference between these practices, for whereas the consumption of meat is far from a modern phenomenon, the consumption of liquid milk in anything but negligible quantities is very substantially modern; it is a product of the modern milk industry. Yet the various problems surrounding milk consumption have been managed or overcome so effectively that drinking milk is now generally more difficult to problematize than eating meat; milk has been 'purified' to a very high degree. Hence the history of the milk industry is remarkably dense with critical possibilities for an exploration of the role of myriad processes of purification in the ongoing constitution of modernity, its conception of 'the social', its deepest assumptions about what it means to be 'human'.

This book deals with a somewhat unusual topic then, and does so in a somewhat unusual way. It fits neither neatly nor easily into established patterns of social research and established genres of social science writing. Nor does all of its subject matter fall within what would normally be thought of as social science. In what it brings within its purview, it may initially strike the reader as slightly eccentric, for at times it marries sweeping narrative with quite particular historical detail, empirical analysis with abstract theory, or complex cultural argument with concrete social history. This is a consequence of two choices I made in determining how to undertake the research for the book and develop the analysis. First, I chose to follow my object of interest as it were 'blindly', to pursue milk wherever it took me. This was a matter of obligation, of feeling compelled to respect the materiality and irreducible particularity of the thing. Hence rather than merely scrutinizing milk in its social dimensions, cultural significations, technological mediation, spatial organization, political significance or economic history, I chose instead to trace the complex historical trajectory of milk as it wove in and out of these domains with striking promiscuity, being here social, there political, here technological, there spatial, and still elsewhere cultural. Adopting this method meant abandoning the security of well-established modes of social scientific understanding, analysis and explanation. But what I found as a result – and what I believe the reader will find in the following chapters – was that far from these various categories (of 'social', 'cultural', 'political', 'economic', and so on), with their associated analytic frames, helping to *explain* the phenomena in question, it was quite the reverse. As I traced the hybrid networks inscribed by dairy milk, I witnessed how these domains were actually being continuously constituted and reconstituted by the diverse associations it forged. It would therefore not do to use the tried and tested analytical resources of 'the social' or even 'the cultural' to explain the heterogeneous connections in question, as the constitution of such domains was a key part of what needed to be explained.

My second fundamental choice was in part a corollary of the first. I chose to adopt the ontological standpoint of generalized symmetry, to suspend all judgement on how things or beings-in-the-world may be separated into fundamentally distinct categories of either 'human' or 'non-human', and how these different *kinds* and their inter-relations may or may not be understood. So, for example, I deliberately wanted to avoid setting out with preconceptions concerning the difference between 'social' things (or 'subjects') and 'natural' things (or 'objects') and how these may properly be known and analyzed. It is a central theme of the book that this great divide is not something fixed or given, but is perpetually being actively reconstituted, or 'enacted', within multiple specific processes, micro-techniques, ad-hoc strategies and local assemblages. My task has been to grasp these processes with as little recourse as possible to foundational ontological categories. In part, this was a matter of conviction – I am philosophically committed to the notion that everything exists and takes place on the same 'level', so to speak, that the world is populated not by classes of things ontologically distinct from each other in *a priori* ways, but by radically equivalent things separated only by particular forms of organization. More specifically, I was (and remain) convinced by the view – articulated most notably by actor network theory – that modernity as a form of ontological ordering involves the separation (and subsequent recombination) of phenomena which are not in themselves separate but intricately connected.

These choices have significantly shaped my approach and they account for much of the unusual nature of the book as a work of social science. Fortunately the very things that make the book unusual will also I think make it highly topical. While belonging to no single discipline or field of research, the approach is such that it should contribute to currently pressing issues and debates across a range of disciplines, from social anthropology to cultural geography. By exploring themes of hybridity, culture–nature interconstitution, heterogeneous materialities and processual ontology, the book is aligned with recent developments in a variety of interconnected fields and traditions, including actor network theory, hybrid geographies, posthumanism, science and technology studies, governmentality and material culture, as well as the late sociological turn to theoretically innovative empirical studies of complex practices interweaving culture, economy and the social. In this sense, it is perhaps best understood as a critique of modernity worked through milk itself.

Acknowledgements

Thanks are due to Mike Savage for continuing to believe in me even at my most idiosyncratic; to Alan Warde for his guidance; to Penny Harvey and Mike Michael for their helpful comments, suggestions and encouragement; and to the 'Culture, Economy and the Social' series editors at the Centre for Research on Socio-Cultural Change (CRESC), without whom this book might never have made it to publication. My intellectual debts are too numerous to mention, but particular acknowledgements are owed to Bruno Latour, John Law, and Timothy Mitchell. Thanks to Aarhus University Press and Wiley-Blackwell for permission to reproduce material in Chapters 4 and 5 which was previously published in *Distinktion: Scandinavian Journal of Social Theory*, 2008, 16, 77–99, and *Historical Sociology*, 2008, 21 (2/3): 272–302, respectively. Thank you to the staff of the Royal Agricultural Society library at Stoneleigh Park, for making me so welcome and for allowing me to make camp in their wonderful archive. Finally, thanks to David Inglis at the University of Aberdeen for his unwavering support and for being a friend.

Introduction

Modernity, humanity and nonhumans

> Man is only a recent invention, a figure not yet two centuries old, a new wrinkle in our knowledge, and he will disappear again as soon as that knowledge has discovered a new form.
>
> (Foucault, 1970, p. xxiii)

> Modernity is often defined in terms of humanism, either as a way of saluting the birth of 'man' or as a way of announcing his death. But this habit itself is modern, because it remains asymmetrical. It overlooks the simultaneous birth of 'nonhumanity' – things, or objects, or beasts – and the equally strange beginning of a crossed-out God, relegated to the sidelines.
>
> (Latour, 1993, p. 13)

What is it to be human? This is not an exclusively modern question, but it is a question that modernity has both asked and answered in a particular way, and its answer permeates the age; it is the basis of the form of order which defines the modern world. So what is modernity's answer? The response is necessarily complex, and its proper unravelling comprises a significant part of what follows. But we must have some starting point, some minimal heuristic sense of how the organization of modernity involves and invokes a specific humanity. I think one can say, then, that in the most fundamental way modernity is human-centred; that is, to be modern is to have an anthropocentric view of the universe, of existence itself. For modernity, 'man' is truly the measure of all things. This is not to say that modern life is incompatible with religious belief, for clearly such belief continues to proliferate in modern societies, yet modernity in principle is profoundly secular, in the sense that it has incorporated humanism into its very organization, by dispensing with the necessity of any reliance upon an external 'God' as the ultimate source of authority for its form of order. Modernity then can be equated with the predominance of secular humanism, with the belief that there is no higher authority than human beings; this is its organizing principle. God may have been crossed out, however, but 'he' has not been obliterated, for humanity has become its own God; at the same time as God has been brought down to earth, humans have been deified. The religious mode of thought has not been dissolved therefore, but has merely assumed a secular form: our age belongs to Feuerbach rather than Nietzsche. From this point of view, modernity is in essence a secular theology, for the fundamental religious cosmology of a centred and meaningful universe remains, albeit with humans as its new centre. Just as God became 'man' in Christian theology, so 'man' has become God in modern secular humanism. In this sense, humanist discourse is quintessentially theistic; its form is theological, even while its content is secular. The banishment of a supernatural god then does not in itself constitute the negation of religious belief, for theological modes of understanding the world persist in multiple forms, most crucially in the modern conception of what it is to be human.

What are the implications of this secular theology, this humanist modernity? As I have said, one is that as God has been deleted, 'man' has been rendered divine. In a dialectical sleight of hand, this has enabled the theological notion that humans are creatures with 'souls' to survive in a secular age. As modern humanists we may not believe that we will live again in a world beyond the grave, or be answerable to a supernatural God, but in a deeper sense we have managed to retain the theological essence of these beliefs, albeit in a secular guise; for we still live as though we believe that humans are somehow 'special', even 'sacred', in a way that both separates us from and elevates us above the nonhuman universe. This universe may be without meaning, a pure contingency, but human life and human action are the exception, we implicitly yet obstinately maintain. Thus our theism lives on in our human exceptionalism, which conditions and underpins almost everything we do in the name of modernity. Indeed, modernity as a form of order, as an ontology, and as an epochal mode of organization, is predicated upon this theistic architecture, this human exceptionalism,[1] without which such defining modern notions as 'progress', 'rationality', and 'truth' would lose much of their effectiveness, if not their very coherence: modernity demands that humans be defined against a nonhuman other. This othering is the condition of possibility of humanist ontology; it is the principle of essential difference which makes humanism possible. But how are humans different, and what are we different from? At the most abstract level the answer is to be found in the subject–object distinction: humans are subjects, while nonhumans are objects, we believe, and from this fundamental difference all else follows. Though this formulation is characteristically modern in its most emphatic and absolute forms, it is not exclusively so, for it has resonated throughout the history of Western thought since ancient Greece, and it still overwhelmingly defines the horizons of possibility of our understanding. We struggle to imagine but that this duality is simply there in the world, given for all time in the order of things and irrevocable; it is the fissure from which modernity is perpetually reborn.

In its Cartesian form it is mind–body dualism, the notion that humans possess 'minds' which are of a fundamentally different order from the materiality of our bodies, yet no mere category of our language. In effect, the 'mind' is posited as a kind of immaterial substance, which is an inescapably theological notion, identical in its ontological form to the idea of the human 'spirit' or 'soul'. This 'mind' is conceived as the seat of human subjectivity, hence of inter-subjectivity and social meaning, indeed of 'social relations' themselves; its separation from the body is the basis of our self-separation from the world of matter, objects, and nature; it is the organizing centre of modernity. We think therefore we are. But the mind does not 'exist' as such; if we cut into our skulls we find only brains, which is to say, more body, more matter, and more nature. 'Mind' is a category of our language, a performative notion; it is enacted into being in our lived practice, which is always already embodied. Or as Deleuze might say, the 'mind' is an *idea* of the body.[2] By positing 'mind' as an immaterial substance, subject–object dualism does two things; it spiritualizes human life and practice, while at the same time reifying materiality by separating it from this life and practice, rendering it inert and abstract. This inert materiality then becomes 'objective' in a Cartesian sense, meaning that our knowledge of it takes a systematically unreflexive form; it is envisaged ontologically as direct and unmediated, hence our active role in its production is made absent. This is the role of dualism in engendering the undialectical 'objectivity' of the modern natural sciences. The counterpart of this Cartesian objectivity is a Cartesian subjectivity, namely 'culture' itself, which is Cartesian 'mind' collectivized. Its corresponding form of knowledge-practice is the knowledge of the social sciences, the doctrine of the autonomy of which parallels the autonomy of 'culture' from 'nature'. Thus the forms of knowledge which modernity both defines and is in turn defined by, are rooted in an ontological dualism emergent from a humanist discourse which is itself traceable to theological principles.

According to this discourse, human beings inhabit an ontologically unique realm qualitatively distinct from and incommensurable with the nonhuman world; this domain is what modernity knows as 'culture'. My central argument in what follows is that this category of 'culture' at the heart of modernity and its knowledge-practices, far from being the given condition of human beings, has been *historically* constituted through the discourse of humanism and the material and epistemic practices in which its categories are actualized. At the core of this is the attempt to show that what we think of and experience as 'culture', which is to say, 'culture' as a mode of organization of knowledge and experience, is produced through a set of historical practices inscribing what it is to be human. My presupposition in making this claim is that the historically specific ways in which we categorize experience and classify knowledge lead to historically specific divisions of experience itself; experience is not an unmediated phenomenon. The corollary is that there is no culture either separate from or preceding 'culture'; culture is not some moment of absolute presence and immediacy. Until quite recently in the social sciences the terms 'culture' and 'human' were invariably used in a manner that was utterly tautological; to be human was to be a cultural being, just as it was taken for granted that to be part of the world of 'culture' was to be human. The conceptual slippage went unrecognized since the two terms were regarded as synonymous. Where the term 'social' was preferred, the same basic assumption was present; humans were social beings, hence to be social was to be human. Modern social science, in other words, is very much the creature of humanist ontology; its conception of its own object of knowledge, which is to say its notion of the 'cultural', and indeed the 'social', has been profoundly shaped by and still largely remains in thrall to the discourse of humanism.

Though the antecedents of this humanism clearly stretch back into antiquity, the mid-nineteenth century saw the beginnings of a significant restructuring of humanist discourse, involving its material realization in a multitude of practices and institutions, its operationalization in scientific and technical knowledges. Not until the later part of the nineteenth century did humanist discourse begin to become materially invested within a whole complex of everyday practices and technologies; rather than a reflective abstraction, the 'man' of modern humanism became the concrete production of an elaborate technico-epistemic apparatus. Now periodization is necessarily inexact, and it projects a linearity onto history which can often obscure more than it reveals; there are wildly divergent periodizations of modernity, for example, some suggesting that it dates back to the sixteenth century, others merely to the early nineteenth or later. I do not view such divergence as a sign of the fruitfulness of the exercise, but of its ultimate sterility. It is finally the relative importance of different dimensions of modernity and the relative merits of different ways of understanding the modern that we are disagreeing about. Casting this in terms of an interminable exercise in periodization strikes me as unproductive, doing little to transform our understanding. For this reason I do not want to make any particular contribution to historical periodization a central part of my argument. On the other hand, periodization seems to be inescapable at some level, insofar as it is both implied by the argument I want to make and necessary as a means to situate it. I will therefore say that I am in broad agreement with Foucault when he posits

> two great discontinuities in the *episteme* of Western culture: the first inaugurates the Classical age (roughly halfway through the seventeenth century) and the second, at the beginning of the nineteenth century, marks the beginning of the modern age. The order on the basis of which we think today does not have the same mode of being as that of the Classical thinkers.[3]

In my view, this shift was material, institutional and technical as well as epistemic; it marked the beginning of a widespread operationalization of humanist categories.[4] It is this ongoing process of practical and material realization which is the object of my analysis in what follows. I want to show how a conception of the human being which is ultimately theological in nature is micro-engineered into even our most technical and scientific knowledge-practices.

In several respects, my view of modernity as so far expressed broadly parallels the ideas of Bruno Latour, actor network theory, and science and technology studies. These traditions have developed the notion of 'generalized symmetry' as a method for unravelling the dualist double-knot of modernity and revealing the ontological politics of its form of order, by undertaking to explain 'social' and 'natural', 'human' and 'nonhuman' phenomena in radically similar terms. From this point of view, the separation of 'culture' from 'nature' emerges not as an ontological given, but as the product of modernity's ongoing 'work of purification', which perpetually inscribes this dichotomy upon the world as a condition of its own coherence and authority. There is much that I admire in this approach and I certainly draw upon aspects of it, but I also depart from it in a number of respects. These departures are complex and progressive throughout what follows and I do not want to present them in truncated form here, so I shall restrict myself to briefly outlining only what has shaped not just my argument but also my choice of historical case study: it turns upon the special significance of nonhuman animals. In short, I contend that Latour's conception of 'the modern constitution', and indeed actor network theory more widely, tends to mobilize an abstracted conception of relationality which obscures the specificity of particular nonhumans. It leads, for example, to a neglect of the particular role of nonhuman animals in the ontological organization of modernity. By attributing primacy to relationality itself (as the word 'network' and its various alternatives imply) and hence conceiving relations as generative *of* rather than generated *by* the things-in-relation, it becomes impossible to view human–animal relations *as* relations in their own right, inter-corporeal relations which powerfully challenge our humanist conceptions of relationality. Instead they are grasped in quasi-semiotic terms as the second-order inscriptions of a modern ontology centred upon a subject/object, society/nature dualism; thus 'things', and especially technological objects, are allowed to take precedence over 'beasts' in defining the domain of nonhumanity.[5] As a result, the irreducibility of animals as a unique kind of nonhuman is neglected, and their subsumption under the object/nature pole of the modern constitution is finally perpetuated rather than challenged.

My position turns this on its head by positing human–animal relations as uniquely critical for modern ontological dualism, suggesting that our actual, lived relations and encounters with nonhuman animals are the most perpetually problematic, unstable and marginal case for modernity's work of purification, and that the government of these nonhumans plays a key role in the constitution of modern 'culture'. This is not a form of base and superstructure causality, with human–animal relations somehow structurally determining the shape of modern ontology. Rather, human–animal relations are identified as an ever-present contradiction lurking on the margins of the categories of humanist discourse, which must therefore be perpetually effaced. It is the necessity of this constant process of containment and deletion that effectively renders animal relations the absent and unstable centre of modern dualism, and their effacement a key enabling condition of modern 'culture'. That is why I prefer to use the term 'humanism' rather than 'the modern constitution' or some similar phrase, for, not unlike Derrida, I wish to emphasize the *species* relations underpinning the dualism of modern ontology.[6] This is critical because our encounters and relations with nonhuman animal others, unless meticulously policed by networks of humanist discourse-practices, have the potential to induce destabilizing and transformative reflections upon our own 'nature' as humans in a way that is very different from our encounters with technology. It is not 'cyborgs' then which are at the frontier of our encounter with a possible post-human other, but 'beasts', 'mutants' and 'monsters'. Thus the typical focus upon technological nonhumans can best be displaced by emphasizing the critical importance of living nonhumans, which may well have been modified by human agency but are not in any primary sense human 'constructs'. I develop this argument more fully later, so I will not elaborate upon it further here. I mention it primarily in order to explain my choice of historical case study, which might otherwise seem arbitrary in relation to my argument.

My case study is the modern British dairy industry, and the liquid milk trade in particular, from the origins of its transformation into a modern industry in 1865, through to the dramatic commercializations of the 1920s and 1930s. I do not use it to 'test' my 'theoretical' argument as such; nor for that matter is the case study a mere illustration of 'the theory', though aspects of the analysis are generalizable. Rather, the case study is integral to the development of the argument. Indeed, my historical analysis and my 'theoretical' argument have advanced simultaneously and organically; they have been largely inseparable in my thinking and I have generally tried to write the book in a way that reflects this. The exceptions are the first chapter, which is a general critique of humanist dualism in social thought and is mainly theoretical, and the conclusion, which is also largely theoretical; whereas historical and theoretical analysis are intertwined in each of the intervening chapters.

The case study is conceived as an analysis of humanist discourse and practice in the British milk industry of the late nineteenth and early twentieth centuries. The historical thesis informing this analysis is that the period from 1865 to around 1939 in dairy agriculture saw an unprecedented entrenchment and actualization of humanism, through the web of disciplinary technologies which were deployed under the rubric of sanitary discourse and practice and the nascent sciences of veterinary pathology, microbiology, and scientific breeding and nutrition. I argue that this realization of humanist discourse was intimately connected to the struggle against diseases of cattle, to the expansion of the milk trade and the associated drive for clean milk, and crucially to the battle against contagion transmissible through dairy milk. I contend, moreover, that this process was closely bound up with the rationalization of practices and institutions constitutive of 'late' modernity or modernity proper. Though I suggest that their common matrix was a modernity constituted in and through humanist discourse, I do not try to address the question of the relations between the field of dairy agriculture and other fields of practice in any systematic way here, not least because it is a deeply complex historical problem and largely beyond the scope of this research. The argument I do make is grounded in a critical discourse analysis of documentary archive sources triangulated with consultation of secondary source materials.

My key primary sources consist of two major national agricultural journals, namely the *Journal of the Royal Agricultural Society of England* (RASE) and the *Journal of the British Dairy Farmers' Association* (BDFA), each of which I have studied in detail for the period in question. While these were key journals of the dairying community, their content cannot be held to constitute anything approaching a scientific cross-section of views from what was a very heterogeneous industry. Though they do contain articles from 'ordinary' farmers, these were generally among the more educated and 'progressive'. The sources cannot therefore be taken at face value as providing the basis for a faithful and 'representative' sociology of all sections of the dairying community. They do, however, provide an excellent basis for a critical analysis, if read carefully and 'against the grain', and that is much closer to what is attempted here. Other sources utilized include the *Journal of the Board of Agriculture*, the reports of the Medical Officers of Health for Manchester and Salford, and a wide range of occasional papers, pamphlets and reports on issues around milk hygiene and the milk supply, produced by local and national authorities as well as sanitary societies. A comprehensive list of sources is included in the bibliography. As well as utilizing these empirically for the historical information they contain concerning changing dairy practices, I have also sought to trace and to problematize the organizing categories and structures of knowledge which they mediate, and to situate these materially in relation to contemporary knowledge-practices in order to grasp their role in the historical constitution of subjects, objects and domains; in this respect my methodological approach is that of genealogy.[7] I will say more about this in the course of the subsequent chapters.

In terms of the practical aspects of my method, it has been an exercise in documentary archive historiography of the kind that is bread and butter to historians; I do not want to render it 'scientific'. In deploying this method I have followed no strictly logical or quantitative sampling procedure, but have selected documents for use on a case-by-case basis and on grounds of relevance, as well as on grounds of availability; these sources have been triangulated where possible. Nor is my approach really 'ethnographic' as such, and it is not intended to be; there are both practical and methodological reasons for this: At a practical level, my choice of case study probes the historical limits of ethnographic analysis, properly so called. It is beyond the reach of living memory, so an oral history is out of the question. Equally, 'ordinary' dairy farmers of the late nineteenth and early twentieth centuries did not spend their time compiling written records of their experiences – they were far too busy working very long hours to produce this kind of 'culture'. What few records there are consist of the official milk records discussed in Chapter 5, or they tend to derive from the more affluent gentlemen-farmers, whose views are already well represented in the sources I have examined. So a documentary ethnography in the sense of a 'history from below' is ruled out. For methodological reasons as well I wanted to avoid an ethnographic approach, despite the popularity and status of ethnography within 'qualitative' social science, because I wanted to explore how human–animal relations are enacted within heterogeneous networks far wider and deeper, and yet at once less tangible and more durable, than direct human–animal interactions. I felt that a closely ethnographic reliance upon farmers' accounts of their relations with animals would be tantamount to re-inscribing the human subject as the centre of the 'social meaning' of the human–nonhuman relations in question, at the level of my methodology. My aim was not to focus upon how farmers' perceptions of their animals shifted over time, which would have been a much more conventional ethnographic-sociological approach, but was rather to explore how 'culture' is inscribed vis-à-vis an animal other, through practices, knowledges, disciplines and technologies which are not only over and above the farmer-as-knowing-subject, but which micro-constitute that very 'subject' as an epistemic entity. In short, my feeling was that the discourse of ethnography broadly speaking is methodologically humanist or inter-subjectivist in its assumptions in a way that is difficult to reconcile with a radically symmetrical approach.

Although I have taken the historical component of the research seriously, it has not been allowed to structure the book as such, in that the chapters are not strictly chronological. Neither, for that matter, are they organized purely thematically. Instead each chapter explores certain themes and certain dimensions of the general argument, and does so by working through not only a specific period in the history of the British dairy industry, but also by focussing upon particular aspects of the industry. As a result, some aspects of the industry may be studied in detail in one period but only touched upon in another; while equally, some dimensions of the argument may be pursued in depth with regards to certain aspects of the dairy industry, but hardly elaborated with regards to other aspects; finally, certain aspects of the industry may be closely analysed in one historical period while other aspects are emphasized in another period. The whole that emerges is therefore something of a patchwork quilt, rather than a strictly systematic whole; there is some overlap, and there are some 'gaps'. There is no aspiration to provide an exhaustive or complete account of anything, and I do not regard this as a weakness.

Chapter 1 then is predominantly theoretical, undertaking a critique of human–nonhuman dualism in social thought. Such a critique might be thought in some quarters to have been rendered unnecessary by recent developments, but this would be to generalize wildly from what is still a relatively marginal current in social science. On a broader view, the case against dualism still very much needs to be made. The strategic emphasis here is upon the classical foundations of sociology, with considerable attention being given to Marx, Durkheim and Weber. In this way the historical roots of the long-standing opposition between constructivism and realism are traced, and the role of humanist discourse in producing this dichotomy is explored. The chapter then discusses actor network theory, arguing that its 'symmetrical' account of socio-material relations represents a significant advance on earlier 'dialectical' theories of society–nature relations. It extends this 'symmetrical' approach by arguing that the central organizing concept of the social sciences, namely 'culture' itself, is essentially the discursive product of human–nonhuman dualism, and that this has profoundly conditioned not only the possibilities of social thought, but also the possibilities of lived social order. Finally, the chapter tries to show why an analysis of human–animal relations is a powerful critical tool with which to problematize and undermine this whole way of knowing.

Chapter 2 introduces the historical case study, outlining some of the key developments in the British dairy industry from 1865 to 1930, and tracing its transformation from a series of localized and informal production-consumption networks into a highly rationalized national industry. An analytical dimension is provided through a focus upon the spatial and temporal dynamics of milk production and distribution networks, and by a critical examination of the political technologies deployed in the milk advertising of the 1920s and 1930s. The aim of the chapter is twofold: on the one hand, it provides the necessary socio-economic and historical context of the case study, but it also serves to raise critical questions as to the ontological conditions of commodification and modernization, and about the role of humanist discourse in creating those conditions. These questions are then pursued in subsequent chapters. The chapter concludes by arguing that the structural political and economic shifts traced here were predicated upon prior ontological and epistemic transformations, and that a detailed genealogical approach rather than a more conventional historical sociology is needed in order to grasp those kinds of shifts.

Chapter 3 is organized into three parts: The first considers the ontological significance of zoonotic disease and the nature of its threat to the humanist categories of modernity. It also clarifies my understanding of humanist discourse, its instruments and disciplinary technologies. The second part commences the properly historical and empirical substance of the chapter. It comprises an analysis of the complex historical conjuncture in which agricultural statistics and experimental microbiology gave rise to the aggregation and politicization of animal disease in the wake of the 1865 'rinderpest' or cattle plague, and which laid the ground for the interventions of sanitary discourse and veterinary pathology. The third section contrasts the fluid and hybrid spatiality of the zoonose with the analytical space produced by sanitary science and practice, which sought to re-impose the dualist categories of humanism upon a disordered zoonotic space. It considers the progress of sanitary legislation against robust opposition, and crucially how the tension between sanitary imperatives and economic exigencies was rooted in their contradictory principles for the use of space. I conclude by bringing together the key developments and by suggesting how the opening arguments may be borne out by the intricacies of actual history.

Chapter 4 builds upon the previous one by drawing more deeply upon the analytical frameworks of 'governmentality' and 'generalized symmetry', associated with the work of Foucault and actor network theory respectively. The aim is to show how these approaches offer complementary strategies for grasping the constitution of the humanist 'subject' and the modern 'social' through micro-political technologies. In particular, it tries to show that the ontological separation of human from nonhuman and culture from nature is enacted *within* such technologies, which can therefore be seen as *ontological* technologies. Rather than remaining an abstract theoretical discussion, this agenda is worked through empirical case materials. This chapter therefore examines the period from around 1890 to 1920 in dairy agriculture which saw an unprecedented realization and concretization of humanist discourse through an emergent network of political technologies. Using historical documentary analysis, it traces the development of several of these technologies, showing how the progressive but uneven realization of a particular kind of 'social' was intimately connected to the sanitary drive for clean milk in the towns and cities, and crucially to the battle against zoonotic tuberculosis transmissible through dairy milk.

Finally, Chapter 5 examines the development of the practice of milk recording in the British dairy industry in the early decades of the twentieth century, focussing upon the analysis of recording as a process of rationalization, in which the formal measurability of milk yields acted as a powerful mechanism for the transformation of dairy production. The chapter briefly outlines the origins of milk recording and explores the social relations which it articulated, the nature and extent of resistance to the practice, and how it underpinned the wider restructuring of the industry. It then goes beyond this historical-sociological account by arguing that within the technical, social, and political processes of rationalization, another less tangible kind of historical process was at work, involving a 'purification' of the human–nonhuman divide. The chapter explores the process of rationalization, via milk recording in particular, as a disciplinary mechanism through which this work of purification is achieved. It argues that milk recording constitutes a humanist technology, operating within and through the pursuit of economic efficiency.

The concluding chapter draws out the critical implications of the preceding analysis, but it is also an auto-critique, which turns upon and problematizes some of the underlying assumptions of the foregoing chapters. It acknowledges that the 'symmetrical' approach of actor network theory has been central to social scientific attempts to acknowledge and respond to the crisis of modern 'social' ontology associated with developments such as climate change, bio-technology and genomics; but it argues that actor network theory has not always lived up to its own calls for symmetry. This chapter therefore undertakes an immanent critique of actor network theory, arguing that its conception of materiality inadvertently reproduces the formalism of the semiotics from which it takes much of its logic. Such formalism has allowed the specificity of certain kinds of nonhuman to be neglected in actor network theory's approach to human–nonhuman relations. This is most pronounced in its preoccupation with nonhumans produced by humans, hence with technologies, which has been at the expense of any significant engagement with nonhuman animals. Attempting to move beyond this impasse, this chapter explores the implications of a strategic decision to reverse this asymmetry, reflecting upon the possibilities of a rapprochement or cross-fertilization between post-positivist social science and an emerging paradigm of cultural ethology.

1 The anthropocentrism of 'culture'

A critique of humanist discourse

Introduction: 'culture' in question

> The definition of culture is vital to notions of the objects of study, the methods and aims of a range of academic disciplines.
>
> (Brooker, 1999, p. 51)

The social sciences begin from the fact of culture; they do not explain it to us. Though cultural sociology, for example, defines itself centrally in terms of the study of culture, it tends to treat culture as an ontological given, as the condition of human existence, rather than as a contingent historical, discursive, epistemic category; as culture rather than 'culture'. Thus culture is extolled as the prior enabling condition of socio-anthropological knowledge. What I want to suggest, however, is that this epistemic formation – indeed the whole modern episteme of which the human sciences are an active element – is actually deeply implicated in the constitution and reconstitution of the ontological vision of the world bound up with 'culture'. Why does this matter? It matters because this approach to 'culture' has rendered social science insensible to the ways in which this category, the ontological organization of the world to which it is integral and the practices in which it is realized, have profoundly conditioned and structured not merely the possibilities of social thought, but also the possibilities of lived social order. My aim in this chapter, therefore, is to undertake a critical analysis of the work of the ontological architecture of 'culture' in shaping the social scientific project and sociology in particular. This will not be my only target, however, for it would make little sense to single out 'culture' for critical treatment without doing the same for 'society'. Of course, 'culture' and 'society' have somewhat different etymologies and have by no means had identical semantic or theoretical histories. But it is possible to remain sensitive to these specific etymological trajectories while recognizing that at the level of the fundamental ontological framework of modern social science, 'culture' and 'society' have not only become increasingly interchangeable in the wake of the cultural turn, but were always residually so in the logic of their othering of 'nature'. Perhaps counter-intuitively, this is underlined by the fact that these terms have often been deployed as the banners of opposed theoretical and methodological positions, and used as rubrics for conflicting conceptions of the social/cultural. For in these controversies, though 'culture' and 'society' often claimed to refer to very different kinds of domain, only their basic underlying identity as ways to define the human existence against nature allowed the terms to be used so effectively in this way, that is, to demarcate opposing visions of the proper object of social scientific knowledge.

The logic of a critique of the anthropocentrism of social theory via a critical analysis of the category of 'culture' might well be thought questionable. Is it possible to speak of a homogeneous 'social theory', not to mention to undertake a general critique of it? Well, no, of course not. Even to take sociology, the discipline which will be my main focus, as though it were a homogeneous entity, would already be absurd, given the rich plurality of traditions that have developed in its name. Fortunately my aim is a more limited and strategic one, in that I want to offer an analytical account of the role of the conceptual apparatus of 'culture' in the development of sociological thought. I have necessarily been very selective and do not aspire to furnish anything resembling an exhaustive account, nor is my argument necessarily exclusive of other narratives. I merely aim to contribute to the opening up of new avenues of critical possibility. While the development of a discipline such as sociology is characteristically anarchic,[1] even fickle, I nevertheless believe it possible, indeed necessary, to identify common threads running through the various sociological traditions in the form of paradigmatic assumptions and frames of reference, the deep epistemological building blocks on the basis of which the discipline has developed, albeit in very divergent directions. It may seem odd in this respect to privilege Marx as much as I do in this chapter, given that he was never a sociologist as such and that Marxism occupied a largely external and oppositional role in relation to the development of academic sociology, at least until the 1970s.[2] Nevertheless, the influence of Marx's writings in shaping the conceptual possibilities of subsequent social thought should not be underestimated, even in those instances where there is no explicit identification with any of the traditions of Marxism. In particular, the seeds of what became the tenets of Weber's and Durkheim's sociologies were clearly planted by Marx. While acknowledging then that this can only be a provisional and controversial judgement, I believe that Marx stands alone among the classical social theorists, both in terms of his overall influence upon later social thought and, more importantly, in terms of his dialectical position on the nature/culture nexus. That a basic dualism structures Weber's and Durkheim's positions on nature–society relations is more or less transparent and explicit; the same can by no means be said of Marx, whose dialectical approach remained the nearest thing to a non-dualist theory of nature/society until relatively recently with the emergence of actor network theory, the turn to space and practice, and the new materialism. As Timothy Mitchell has argued:

> In social theory there is an important exception to the rule that human action is put at the centre and the external world is treated as an arena for such action rather than the source of forms of agency and power. It is found in the work of Marx.
>
> (2002, p. 30)

Moreover, the very notion of modernity suggests that there is something inescapably common to classical and contemporary social thought, and Marx was the most brilliant and seminal analyst of modernity. That is why at some level Marx and modernity continue to be inescapable for contemporary social theory. Even where it consciously defines itself against modernity and claims boldly that 'we have never been modern', it finds itself defined by modernity nevertheless in the inner dialectic of this very manoeuvre. One might say that social theory makes its own history, but that it does not make it just as it pleases, but within discursive and epistemic formations directly encountered, given, and transmitted from the past.

The others of 'culture' and 'society'

In what does the category of 'culture' consist then? Without doubt the word is polysemic and has a long history prior to its adoption by social science; indeed it is generally regarded as probably the second most complex word in the English language, the most complex of all being 'nature'.[3] Derived from the Latin 'cultura', it first appeared in English around 1440; among its earliest meanings are 'organic cultivation', 'tillage', 'husbandry', or the care and nurturance of plants and animals, the tending of natural growth.[4] In reference to the cultivation of the human mind or body through education or training, 'culture' is first recorded in the works of Sir Thomas More about 1510. It was not used in the sense of the intellectual or artistic dimension of civilization until 1805, in Wordsworth's *The Prelude*. In its origins then, far from denoting nature's immaterial other, or distinguishing human activity from nature, 'culture' is a word derived from the exigencies of material existence, from agricultural labour, and from the embodied experience of interaction with nonhumans. The historically changing meanings of the word, however, manifest a gradual but definite shift away from this organic sense of corporeality, agriculture and material activity, and towards precisely immateriality, the life of the mind, consciousness and the spirit. This etymological movement was broadly contemporaneous with the emergence and development of modernity, and particularly industrialization and the growth of the towns. It was this distinctly modern version of 'culture' as the opposite of nature that entered into the lexicon of modern social thought, acquiring further theoretical and technical inflections in the process. Indeed, as Terry Eagleton writes, 'The word thus charts within its semantic unfolding humanity's own historic shift from rural to urban existence, pig-farming to Picasso, tilling the soil to splitting the atom.'[5] One of the principal arguments of this book is that 'culture' does not merely chart the emergence of modernity but has contributed actively to this emergence; that is to say, the ontological and epistemological apparatus bound up with the concept of 'culture' has been intrinsic to the proliferation of the knowledge-practices constitutive of modernity as an historical formation.

The etymology of the word 'society' manifests a somewhat different pattern of development. From the Latin 'societās', it first appeared as 'society' in English in 1531 meaning 'companionship.'[6] In its meaning as an organized group or association 'society' is first recorded around 1548, and in the sense of a system or condition of living with others in a community, in 1553. What is remarkable about both 'culture' and 'society' is that the history of their changing meanings reflects not only the process of modernization but also the class-divided character of human experience. Thus 'culture', meaning the tending of natural growth through careful labour, came to be progressively disconnected from this earthy, corporeal association, with the emergence of industrial capitalism and the modern bourgeoisie, coming instead to signify such intellectual and artistic activities as could be pursued only by those with sufficient freedom from daily bodily labour to pursue the cultivation of the mind and senses. This became the dominant meaning of 'culture' during the later nineteenth and early twentieth centuries, and it persists today, as a striking vindication of the thesis that our language 'reflects', or better, is organically inseparable from, our form of life, our practical life activities and conditions of existence. Equally, 'society' had no connotations of exclusivity or distinction in its early meaning as the down-to-earth 'companionship', but again this shifted with time and in 1823 'society' was for the first time used in the sense of the fashionable elite and their activities, by Byron in *Don Juan*.[7] Interestingly, although 'society', unlike 'culture', had no specific organic or agricultural associations at the beginning, it grew to acquire precisely such significations with the development of modernity, as it began to be used as a synonym of 'culture'. In 1899, 'society' was first used in ecology, in the sense of a plant society, or in other words a 'culture', marking the beginning of the next phase in a centuries long process of convergence in the meanings of these words.[8] As 'society' then began to be used interchangeably with 'culture', at first only with reference to the older organic sense of 'culture', it is no surprise that the modern shift in the meaning of 'culture', from cultivation to consciousness, from the dirt to the heavens, in turn exerted a parallel influence upon the meaning of 'society' which would eventually bring it into congruence with the newly immaterialized 'culture'. It is in the social sciences, above all, that the final stages of this dialectic have been played out.

Of course 'culture' is given distinct inflections by different schools of social thought, and these inflections have themselves shifted over time, but this does not lead inexorably to the conclusion that it is illegitimate or impossible to undertake an analytical abstraction from these diverse uses in order to identify 'family resemblances'. The undeniably contested, contradictory, local, overlapping and divergent meanings of the word 'culture' do not preclude the identification of its most persistent ontological assumptions, but are merely the rationale for such an exercise. I take it as read that analysis as a technique presupposes the complexity of the world, far from being foiled by it. At the necessary level of abstraction then, it is clear enough that 'culture' as it has been rendered by social science denotes a system of inter-subjective symbolic meanings, or a 'signifying system'.[9] This conception involves a complex of several distinctions, several others by way of contrast with which the multi-faceted identity of 'culture' emerges. The first of these is 'society', and by extension 'social structure', 'economy' and 'politics', which have often been used as though they were nomenclatures for ontological sectors of reality or distinct 'levels' of phenomena. 'Culture', for its part, specified the 'level' of the production and communication of inter-subjective meanings. In some versions this 'level' was relatively autonomous, while in others it was thought to reflect or represent some bedrock of 'social structure'. This usage was often associated with the notion that 'society' was the totality underpinning or encompassing these ontological domains. With the advance of 'culture' and the cultural turn in social science, these domains have increasingly become conceived as immanent sub-sets of the totality of 'culture' which permeates them; so that, for example, sociologists increasingly regard social structure as culturally constituted rather than separate from 'culture', and social anthropologists now routinely speak of the culturally constructed nature of economic value and the cultural life of commodities. Thus 'society' as structural totality has been succeeded by 'culture' as immanent totality, but the totalizing character of 'culture' has gone largely unacknowledged, due to both its immanent form and its perception as a decentred and heterogeneous category.

The second other is 'matter', which semiotic conceptions of 'culture' construct as its mechanical opposite, purely physical and by definition meaningless until its 'construction' or 'signification' within 'culture'. Hence, for social constructivism, matter has no agency. Against this it might be objected that Raymond Williams' influential definition of 'culture' as 'a whole way of life of a social group or whole society' offers a real and promising alternative to this semiotic vision.[10] But Williams himself argued that his conception was intricately linked to that other notion of 'culture', as intellectual or artistic activity, *in and through* the concept of 'culture' as *a signifying system*, by means of which, necessarily, 'a social order is communicated, reproduced, experienced and explored'.[11] In this view, 'culture' is 'a whole way of life' precisely because it is the signifying system through which this whole emerges *as* a whole. Thus, even if a 'whole way of life' includes such material things as tools and modes of dress, these are only deemed relevant because and insofar as they are symbols of the system of meanings or beliefs that is the 'culture'; it is their semiotic identity that is culturally significant. It turns out then that even Williams' apparently holistic and practical conception of 'culture' is logically predicated upon the semiotic architecture. Moreover, it is the absent presence of this semiotic 'culture' permeating all social scientific uses of the word that has allowed it to be used to refer to 'a whole way of life' in the first place.

The third other of 'culture' is 'biology', which in the shape of a mythical 'biological determinism' has been central to the self-valorization of all shades of constructivist social science, but cultural feminism in particular, for which the refutation of all forms of this determinism became conceived as a matter of political necessity and hence as an article of faith, according to the following logic:

> People do not revolt against what is natural, therefore inevitable; or inevitable, therefore natural. Since what is resistible is not inevitable; what is not inevitable could be otherwise – it is arbitrary therefore social. The logical and necessary implication of women's revolt, like all revolts, is that the situation can be changed. Belief in the possibility of change implies belief in the social origins of the situation.
>
> (Delphy, 1984, p. 211)

What is most striking about this is that political ideology is explicitly the driving force and theory its post-hoc rationalization, while the equation of the 'natural' with the inevitable is its key enabling assumption. As a result of these moves, 'culture' and 'society' become simply terms for what is politically mutable. 'Biology' is not in any way a 'social' determinant, according to this reasoning, but only the manner in which 'biology' is interpreted and given meaning within the 'culture'. The corollary is that an absolute separation is maintained between 'biology' and 'culture', the two domains are held strictly apart in a dichotomy which defines and constitutes each as a separate 'pure' domain, free from contamination by the other; pure biology on the one hand and pure culture/society on the other, as though they really existed separately from each other as anything but hypostatized abstractions.[12] Outside of feminism the logic was very similar, in that it was necessary to maintain that our lives are 'socially constructed' in order to be able to continue to insist upon the possibility of their fundamental transformation. The social revolution was always in the future and on the horizon, from where it could justify the form of theory (as ontological purification) in the present.

Underlying such political commitments was a deeper rationale rooted in the politics of the demarcation of disciplines, upon which the authority of disciplinary knowledge claims has been founded, not to mention countless academic careers.[13] Hence, the effort to establish social constructivism against a biological foe not only denounced this enemy in a highly charged moral and political discourse but simultaneously fixed the identity of 'biology' in terms which were self-validating of social constructivist social science, that is, in terms of a reductive 'biological determinism'. 'Biology', like materiality, was therefore inscribed as the mechanical opposite of 'culture', as a domain characterized by a static and crude form of determinism, without 'social' meanings except those conferred upon it by 'cultural' signification and made visible by social science. For all its implausibility as a way to describe our lived experience and practice, our thoroughly heterogeneous being-in-the-world, which is always already embodied and biological in ways that really matter, this project has had great rhetorical force and strategic effectiveness within the academy. Indeed, it has been so successful that any reference to 'biology' is now suspect within the social science community and must usually be carefully qualified or risk derision.[14] 'Biology' has been effectively censored through a polarity of biology/culture in which only one pole is deemed 'social'; and though there are signs now of its incremental return in sociological theories of the body, this return is strictly 'cultural' and at all times under the conceptual domination of a constructivist orthodoxy, so that the agency of corporeality itself remains silent.

'Culture' then, is not quite the holistic and decentred category it is generally taken to be. On the contrary, 'culture' is both reductive and imperialistic: the latter because it is tendentially totalizing, bringing other categories increasingly under its sway and progressively subsuming their specific ontological textures under its own logic as a closed system. This would not be problematic in itself if the ontological specificity of other categories were retained within the emerging totality of 'culture', but this totalization is also a surreptitious yet systematic process of reduction, for the whole that remains, whether explicitly or otherwise, is a totality of signification, a totality of disembodied inter-subjective meaning. Everything is encompassed by 'culture', nothing categorically excluded, allowing cultural theory to engage freely with materiality, bodies and objects at length, and to render them objects of its knowledge; but in the process they become 'cultural' objects. Hence the apparent inclusiveness of 'culture' conceals a work of ontological filtering or purification, so that the 'matter' which is incorporated is no longer really material, but is transformed into the *meaning* of specific materialities within the signifying system. Materiality itself is thereby made absent by being relegated to the epistemological status of a noumenon, a matter of speculation, forever unknowable. It is even routinely dismissed as a metaphysical concept; and so it really appears to be from the point of view of a conception of 'culture' which is itself metaphysical but does not know it. Hence the attempt to acknowledge actually existing and lived materiality in its heterogeneity appears as though it were an attempt to specify some pure essence, knowable directly, without the mediation of the 'signifying system', which would indeed be a metaphysical and essentialist proposition from this point of view. But this point of view depends upon a prior and unacknowledged separation of 'mind' (rendered collectively as inter-subjective 'meaning') from 'body' (grasped transcendentally as 'matter'); a Cartesian separation which involves a hypostatization of 'the signifying system' via its abstraction from lived materiality and from a practice that is constitutively embodied.

The category that has been most resistant to subsumption under 'culture', at least until very recently, is the concept from which 'culture' is originally derived: 'nature' itself. The distinctions between 'culture' and its aforementioned others are organized around a remarkably persistent dualism of 'culture' and 'nature', which permeates each of these other, subsidiary distinctions. In this sense 'nature' is the final and most fundamental other of 'culture', something like its antonym, against which it has above all been defined; 'culture', in the end, is that which 'nature' is not. Thus 'social structure', 'politics' and 'economics', for example, have at different times been distinguished from 'culture' through what one might call the objectivist fallacy, the notion that certain 'levels' of social phenomena are comparable to object-nature in their apparent externality, obduracy and durability in relation to human consciousness, in other words, their disinclination to bend to the human will. Similarly, 'matter' was linked to 'nature' because it was independent of the 'mind' or 'consciousness' of the human 'subject'. For constructivism, however, this independence became problematic, for it could easily be construed as something like a form of agency attributable to matter itself; if matter possessed agency then it might even be meaningful, rather than merely a meaningless noumenal object onto which inter-subjective meanings are projected. This in turn would undermine the humanist subject, the politico-ontological heart of modernity. The solution to this problem was to erase matter entirely, or rather to cross it out epistemologically by insisting that matter was unknowable except within a semiotically defined 'culture'.

Matter was thereby reduced to a mechanical substrate without 'agency' of its own, beneath 'culture' but never entering into 'culture' except by way of becoming it. This is Cartesian materialism, which is to say the conception of materiality that corresponds to and complements the Cartesian idealism of 'culture'; in its contemporary form it is known as transcendental realism, the epistemology instinctively favoured by natural science.[15] Crucially, this move meant that the acknowledgement of 'material culture' did not mean the acknowledgement of the 'materiality' of culture, for the 'culture' concept itself remained ontologically unchanged by its encounter with matter. In this way, the immateriality of culture was retained even in the innermost logic of its dialectic with matter. The matter in question was strictly outside of 'culture', rather than inextricably bound up with 'culture' in a state of being and practice so constitutively hybrid that it makes a nonsense of their separation. As for 'matter' so too for 'biology'; that organic 'matter' from which our bodies are constituted. Indeed, the association of 'biology' with 'nature' is so deeply entrenched as to have become common sense, according to the following logic: as 'biology' is not under the control of our conscious 'agency', it is other than 'culture', therefore it is 'nature'; if 'biology' is 'nature', then biological agency is natural agency, which is negated by social constructivism, the discourse of 'culture'.

In each of these distinctions between 'culture' and some other, it is the human 'subject' which is at stake and which is really at the elusive centre of these unfolding distinctions, the subject, that is, conceived in terms of 'consciousness', (inter)subjectivity and 'agency'. There is a deep and basic alignment between, on the one hand, 'culture' or 'the social', and on the other hand, the 'human' as the sphere of the 'subject' and 'subjectivity'; so that each distinction between 'culture' and its others invokes an underlying and parallel differentiation between what is human and what is not. Thus the category of 'culture' posits the human 'subject' against assorted nonhuman 'objects', objects, that is, of the human 'consciousness', of human meaning or subjectivity. In this way the discursive production of humanity or the 'subject' is organized around the concept of 'culture' and by extension the concept of the 'social' which it has increasingly colonized; both involve defining the 'subject' as a being of 'agency' vis-à-vis a morphous nonhumanity conceived predominantly in terms of 'nature' inscribed as an 'object'. This multilayered dualism does not, as some dialectical theorists have assumed, emerge from a fundamental and deeply entrenched subject/object dialectic in occidental thought. On the contrary, it is this discursive, epistemic and ontological formation – which I refer to as humanist discourse – that interpellates both the 'subject' and the 'object' in terms of human/nonhuman dualism. This humanist 'subject' is thus the condition of possibility for subject/object dialectics, rather than the other way around; the subject's historical production precedes its abstract theorization. Nor does this humanist discourse exist in a historical vacuum, but rather, the human 'subject' which it produces is not only distinctively (albeit not exclusively) modern, but actively productive of modernity; it is a human 'subject' integral to the coherence of modernity as an episteme and an order of being.

So the ontological positioning of the 'subject' as 'cultural' vis-à-vis the assorted others of 'culture' enacts and produces modernity at once epistemologically and politically. As Bruno Latour has argued, the separation of 'nature' from 'culture/society' underpins and legitimizes a separation of science from politics and 'truth' or knowledge from power, so that both the materiality of power and the political agency of scientific knowledge-practices and technoscientific objects and systems are excluded from recognition in modern political discourse and liberal democratic practice, by being divided into 'pure' science and 'pure' politics.[16] This disarticulation of knowledge and power is a crucial condition of possibility of modern 'governmentality' as an ensemble of specific rationalities of power built around fields of knowledge and strategies for 'the conduct of conduct' and 'the government of things'.[17] More specifically, by dislocating politics as a sphere of human affairs and human agency, from science cast as knowledge of a nature and a materiality defined in opposition to humanity, humanist discourse makes invisible those technologies of power which consist of capacities to instrumentalize space and materiality in order to discipline bodies and regulate conduct. Humanist discourse is therefore indispensable to creating the ontological and epistemic conditions of modern governmentality, but this relationship is mutual, for the political technologies of governmentality are also the means of actualization of humanist discourse, its *modus operandi*. To open up these mechanisms to political contestation by revealing the politics of scientific practice and the 'agency' of technological artefacts has been the defining project of social studies of science. But the mutual purification of science and politics also conceals the ontological politics of a humanist discourse which, though it does not directly control human 'subjects', yet organizes the ways in which it is possible to know oneself as human, hence to *be* human, in modernity. The possibility and the necessity of challenging this aspect of humanism are only just beginning to be understood;[18] it is to this emerging critique that this book is a contribution.

Humanist-naturalism? Marx's dialectic of nature and society

Intrinsic to the category of 'culture' then is a discourse structured around human/nonhuman dualism, which in turn is integral to the politico-ontological and epistemic order of modernity. How then is Marx's dialectical approach to be grasped in relation to this dualism? Marx's historical materialist conception of the 'nature–society' nexus represents a complex and nuanced position which has generated much interesting work by later Marxists and critical theorists on 'nature'.[19] Marx was one of the first to show, in his analyses of commodities and industrial machinery, how objects could become bearers of human social relations. I want to suggest, however, that Marx's articulation of the dialectic contained a deep ontological asymmetry and was therefore inherently ambiguous in such a way as to leave the door very much open to a humanist-constructivist reading. The foundations of this constructivist reading were reinforced by the work of Weber and Durkheim, who unlike Marx of course were direct contributors to the foundation of sociology as an academic discipline. In their radically different ways, Durkheim and Weber developed certain dimensions of Marx's thinking, but only by tearing them out of the dialectical framework to which they were integral in his work. As a result, through Weber – and especially through those who later developed his methodology in a more consistently phenomenological direction, notably Alfred Schutz – the 'social' became more discursive, inter-subjective, and semiotic;[20] while through Durkheim and his realist inheritors, it became more objective, mechanical and deterministic.

These poles and the diametrically opposed ontologies of 'the social' which they offered, were pivotal in the development of sociology, for as they crystallized they defined the horizons of possibility of the ways in which its object of knowledge could be imagined; they ensured that the great multilayered divide between the various shades of positivism/functionalism/realism and interpretivism/interactionism/phenomenology became inescapable. This bifurcation in turn conditioned the ways in which it was possible to read Marx through much of the twentieth century and was therefore influential in the later development of Marxism. Consequently, the Marxist tradition became itself bifurcated along the fault line of its positions on (among other things) 'nature' and epistemology. Thus the subtlety of Marx's original treatment of nature was obscured under the force of the opposition within Marxism between, first, historicists such as Lukács and dialectical materialists like Kautsky, and then later by the parallel division between 'humanist' and 'structuralist' Marxists.[21] The latter antagonism was particularly significant since it corresponded historically with the co-option of Marxism into the academy and into sociology in the 1970s. It makes more than a little sense therefore to return to Marx himself and to look again at the category of 'nature' in his thought.

Marx identified parallels between his own historical materialism and the historical naturalism of that other great Victorian, Charles Darwin, whose work he greatly admired. There is even some evidence that Marx sought to emulate Darwin's natural history in some aspects of his own approach to historical development. Both posited theories of historical development in which some sequence or logic of necessity emerges from a multiplicity of indeterminate events, allowing both writers to propose a mechanism of historical development without either determinism or explicit teleology.[22] The form of their historical explanations was highly similar in this respect. Moreover, Marx wrote approvingly of *The Origin of Species* that 'Darwin's book is very important and serves me as a basis in natural science for the class struggle in history.'[23] He also enthused that 'this is the book which contains the basis in natural history of our view'.[24] Perhaps the strongest and most influential statement of Marx and Darwin's apparent complementarity was that made by Engels at Marx's funeral, where he declared that 'just as Darwin discovered the law of development of organic nature, so Marx discovered the law of development of human history'.[25] As Tom Bottomore notes, however, 'It is highly questionable whether the positivist presumptions attributed by Engels to Marx and Darwin in common fit either scientist at all.'[26] Indeed, even in this account by Engels, the posited relationship is mainly one of analogy, for note the ontology implied by his words: Nature is separate from history, history being human and social; Darwin's account is adequate for nonhumans, but not for humans, since humans are social rather than merely natural; for this reason they are historical beings, and are not subject to the laws of natural history, but to the laws of human history. Positivist this may be, but it is still very much dualist in its conception of the relation between nature and society. Equally, though Marx admired Darwin's form of historical explanation and his naturalistic conception of 'man', not least because it was perceived as a huge fillip to atheism, he was highly cautious when it came to the political applications of Darwin's theory, and he had nothing but derision for social Darwinism proper, which sought to 'explain' history in terms of human biology, population pressure, competition for survival, and the evolution of the species through selection.

'History' then, for Marx, that is to say human 'social' history, is connected to nature through human agency or praxis, through purposive human action upon nature, rather than through Darwinian natural mechanisms or any other kind of volition or agency on the part of nature itself. Nature was regarded by Marx as the reactive object, humanity the active subject. The term for human praxis acting upon nature was *labour*, which Marx conceived as a metabolic interchange between humanity and nature, a process in which humanity (re)creates both itself and 'society':

> Labour is, first of all, a process between man and nature, a process by which man, through his own actions, mediates, regulates and controls the metabolism between himself and nature. He confronts the materials of nature . . . he acts upon external nature and changes it, and in this way he simultaneously changes his own nature.
>
> (Marx, 1976, p. 283)

> Labour is neither culture nor nature but their matrix.
>
> (Young, 1991, p. 399)

Hence labour is the key mediating moment of the subject–object dialectic at the centre of Marx's thought, at once historically, epistemologically and ontologically. Interestingly, this conception of labour is very close to the etymological sense of the word 'culture' as the cultivation of natural resources. For Marx, social history (subject) has emerged out of natural history (object) yet represents a fundamental departure from it; it is tied to natural premises, but has become other than nature.

Yet Marx has often been seen as a naturalistic thinker, and not without some justification, for he insisted that historical materialism alone gave due recognition to the historical significance of the human existence as a species of natural, physical beings compelled to produce the means to satisfy their material needs. This did not mean that humans were wholly natural beings, however, but on the contrary, it was the way in which humans laboured that determined their social-historical being, over and above their natural existence. Nature merely dictated the necessity of labour, whereas the mode of organization of labour and its technological development not only shaped society, but was itself shaped by society. Marx therefore described his position as a unity of naturalism and humanism, as 'humanist naturalism'. The cogency of this dialectical unity, however, is wholly dependent upon the special work performed by the category of 'labour', for in Marx labour both separates society-history *from* nature, while at the same time, in the process and the medium of this very separation, labour also establishes their socially-historically conditioned interaction. Without this mediating concept Marx's ontology would be quite straightforwardly dualist, for it is only the category of labour that stands between Marx and his otherwise transcendental-realist conception of 'nature'. But it is precisely the capacity of 'labour' to perform this dialectical conjuring trick of simultaneous purification-and-mediation while remaining a materialist concept that I want to call into question.

Marx wants to be able to acknowledge that humans are part of nature and that 'man' is a natural being, while also maintaining that humans alone are 'social' and 'historical' beings, thus:

> Man is a directly natural being . . . But man is not only a natural being, he is a human natural being. This means that he is a species being that exists for himself, thus a species being that must confirm and exercise himself as such in his being and knowledge.
>
> (Marx, 1977a, pp. 104–5)

For Marx, it is humanity's capacity to create and recreate its own 'nature' through its action upon 'external' nature that distinguishes human society from the nonhuman world, for it is from this process of human self-authorship in labour that the social character of humans derives. This ontology is coherent enough with reference to the relations between humans and the inert objects of labour, but if humans alone are to be deemed social, then some distinction has to be made between humans and non-human *animals*. As Marx recognizes, the only properly materialist grounds for such a distinction is labour itself; hence he posits the distinction in these terms:

> Men can be distinguished from animals by consciousness, by religion, or anything else you like. They themselves begin to distinguish themselves from animals as soon as they begin to produce their means of subsistence, a step which is conditioned by their physical organization. By producing their means of subsistence men are indirectly producing their actual material life.
>
> (Marx, 1977c, p. 160)

> We are not dealing here with those first instinctive forms of labour which remain on the animal level. An immense interval of time separates the state of things in which a man brings his labour to market for sale as a commodity from the situation when human labour had not yet cast off its first instinctive form. We presuppose labour in a form in which it an exclusively human characteristic.
>
> (Marx, 1976, pp. 283–4)

But what is this form of labour? Indeed how is this distinction between human and animal labour to be sustained? For as Marx recognizes, non-human animals also labour in order to reproduce their conditions of existence:

> It is true that the animal, too, produces. It builds itself a nest, a dwelling, like the bee, the beaver, the ant, etc.
>
> (Marx, 1977a, p. 82)

> A spider conducts operations which resemble those of the weaver, and a bee would put many a human architect to shame by the construction of its honeycomb cells.
>
> (Marx, 1976, p. 284)

Indeed, there is nothing in the empirically observable character of human labour that separates it decisively from comparable animal activity. On the contrary, these labours are radically equivalent when grasped symmetrically, which is to say, in relation to the form of life in question.

Even judged in terms of complexity, there are many forms of human labour which are more repetitive and routine than certain cases of nonhuman labour, just as there are animal labours which manifest the characteristics of highly organized collective activity.[27] But these radical symmetries are deeply incompatible with Marx's whole conception of the unique place of humans within nature, for without the metaphysical architecture of human exceptionalism, the society/nature dichotomy is undermined. In short, the distinction is theoretically necessary to Marx and it *must* be made; perhaps this is why he tacitly turns away from the observable character of the activity and towards its inner, metaphysical essence, its uniquely human 'soul', as follows:

> We presuppose labour in a form in which it is an exclusively human characteristic . . . what distinguishes the worst architect from the best of bees is that the architect builds the cell *in his mind* before he constructs it in wax.
>
> (ibid., pp. 283–84; my italics)

> Conscious vital activity differentiates man immediately from animal vital activity. It is this and this alone that make man a species being.
>
> (Marx, 1977a, p. 82)

> The animal does not enter into 'relations' with anything, it does not enter into any relation at all. For the animal, its relation to others does not exist as a relation. *Consciousness* is, therefore, from the very beginning a *social* product, and remains so as long as men exist at all.
>
> (Marx, 1977c, p. 167; my italics)

But what exactly does this distinction amount to, when rendered in these terms? It is not what it set out to be; for it is not a distinction between the labour of humans and that of animals, but between the *consciousness* of humans and the (un)consciousness of animals, as ostensibly manifest *in* their labour. Though Marx stresses activity, it is the status of human activity as *conscious* activity – conceived in terms of its becoming an object to itself, hence as *self-consciousness* – which does the conceptual work of separating human from animal activity. It is not the activity itself but its Cartesian mental accompaniment which is being invoked in order to legitimize the human–nonhuman distinction. Thus Marx's separation of human from animal labour rests upon what in his own terms is an 'idealist' conception – that of 'consciousness'; humans possess it, animals do not, it is asserted, and in this fact alone the labour of humans is distinct. Human labour is conscious, intentional, designed in advance, or in other words 'cultural', whereas the labour of animals is unconscious, merely instinctive, without purpose, and therefore 'natural'. It follows that the distinction between human and animal labour in Marx is reducible to an *a priori* distinction between humans and animals on the grounds of 'culture' itself.[28]

The concept that forms the mediating core of Marx's dialectic of nature and culture then, turns out to be based upon a *prior* distinction between the two domains (and in this respect reproduces the modality of religious thought); 'labour' presupposes 'culture' and cannot for that reason explain it. To refute this is not to deny that there are significant differences between human and animal labour, but to insist that these differences are differences of scale, of magnitude, and in some cases of complexity, rather than the kind of absolute difference supposed by Marx. Certainly human and animal labours cannot legitimately be distinguished on the grounds of the purely 'mental' (spiritual no less) component which ostensibly accompanies the human activity exclusively. This is purely metaphysical; for where *is* this mental component of labour? Can one point to it? Of course not, because it is inseparable from the activity and rendered purely abstract when conceived in isolation. The intentionality of labour, human and nonhuman alike, must be found within the activity itself; it is manifest in the observable utility of the result of labour, in the organized character of the activity and the purposiveness of the activity, relative – as it must be – to the form of life in question. Thus the reliance upon a hidden, inner essence ('consciousness'), 'in the mind', to distinguish between human and animal labour, is both methodologically unsustainable and (it need hardly be said) neither dialectical nor materialist. Indeed, it is completely incompatible with a great deal of Marx's stated method, and his actual method, on all matters except this distinction between human and animal labour. For example:

> Consciousness can never be anything else but conscious existence, and the existence of men is their actual life-process . . . In direct contrast to German philosophy which descends from heaven to earth, here we ascend from earth to heaven. That is to say, we do not set out from what men say, imagine, conceive, nor from men as narrated, thought of, imagined, conceived, in order to arrive at men in the flesh . . . the phantoms formed in the human brain are also, necessarily, sublimates of their material life process, which is *empirically verifiable* and bound to material premises.
>
> (Marx, 1977c, p. 164)

> The dispute over the reality or non-reality of thinking that is isolated from practice is a purely scholastic question.
>
> (Marx, 1977b, p. 156)

Or as Wittgenstein liked to say, if you wish to understand the game of chess, do not try to perceive what is going on in the minds of the players, simply attend to what is taking place on the board. In other words, practice is quintessentially 'public' in nature; its meaningfulness is not to be found in the 'mind' of the practitioner, but in its orientation towards others and its comprehensibility by others. To act as though its meaning were private and subjective is to deny this. I can identify no reason, or rather, no reason consistent with generalized symmetry, to restrict the application of this principle to human practice while excluding animal practice from the same epistemological treatment.[29]

This approach is by no means entirely absent from Marx; it is present, for example, in his brilliant account of the phenomenon of language, thus: 'Language is practical consciousness that exists also for other men, and *for that reason alone it really exists for me personally as well*; language, like consciousness, only arises from the need, the necessity, of intercourse with other men.' (Marx, 1977c, p. 167) Here Marx recognizes that language, and by extension anything that might be called 'consciousness', is a fundamentally public, collective phenomenon, which both emerges and is inseparable from practical activity; this is a distinctly non-Cartesian materialist approach. Yet Marx unaccountably reverses this approach when it comes to distinguishing human from animal labour, attempting to explain practice in terms of 'inner' consciousness, when he claims to be doing the reverse, descending from heaven to the earth instead of ascending from earth to heaven. So why would Marx, so fiercely committed to materialist explanation, smuggle this Cartesian element into his central concept? The only answer can be that there are simply no other grounds on which the distinction can be sustained. Marx's overwhelming priority is to emphasize that labour is *social*; but in substantiating this sociality, he conflates it with the claim that labour is exclusively or uniquely human. In this way, Marx inadvertently reproduces the category mistake or self-affirming tautology which defines humanist discourse.

Although historical materialism as a broad theory of history begins more or less foundationally from labour, *Capital* is best read as an extended critique of the commodity-form, which by a progressively wider dialectical unfolding takes in the totality of capitalism and the commodification of labour. While the fortunes of historical materialism, read as a positive theory of history, have been in wane for some time, Marx's critique of capital is still very much at the centre of contemporary critiques of economic organization. Indeed, there is no sign that it will be surpassed in the foreseeable future. Moreover, it is not entirely implausible to posit a degree of independence for this critique of capital in relation to the wider tenets of historical materialism, or at least to cast the latter in loosely heuristic terms more amenable to currently popular interpretations of the former.[30] For this reason it may not be enough to show that humanist ontology is built into Marx's concept of labour; I therefore want to show that humanist discourse also structures the concept of fetishism at the heart of Marx's critique of the commodity-form, as classically summed up in the following passage:

> There it is a definite social relation between men that assumes, in their eyes, the fantastic form of a relation between things. In order, therefore, to find an analogy, we must have recourse to the mist-enveloped regions of the religious world. In that world the productions of the human brain appear as independent things endowed with life, and entering into relation both with one another and with the human race. So it is in the world of commodities with the products of men's hands.
>
> (Marx, 1976, p. 436)

Marx strongly implies here and elsewhere that there is something anthropologically strange, perhaps even something fundamentally 'wrong', with the projection of social relations 'between people' onto things:

> Only the conventions of our everyday life make it appear commonplace and ordinary that our social relations of production should assume the shape of things, so that the relations into which people enter in the course of their work appear as the relations of things to one another and of things to people.
>
> (Marx, 2002, p. 91)

Thus, with his notion of commodity-fetishism, Marx powerfully naturalizes the ontological separation of subjects and objects, suggesting that beneath the distortions of capitalist social relations some anthropologically pure state in which subjects and objects are not all mixed up together but instead quite separate, still exists as something like a 'state of nature', to which communism might one day return us. From this point of view, subject–object interpenetration and hybridity are a perversion of social relations, indeed a 'fetish', with all the derogatory religious associations this word held for Marx. The language used to describe this fetishism is a language of 'spectres', 'ghosts' and 'phantoms', which powerfully underlines what for Marx was the profoundly unnatural status of 'social relations between objects', which he compares to religious illusion in their manner of being unreal yet emergent from real conditions.[31] From the point of view of a symmetrical critique, however, it is the notion of a pure state in which subjects and objects are *not* mutually constitutive which is religious, revealing the theological contours of modern ontology.

This discursive purification of the eclectic heterogeneity of actual practice enables the distinction between nature and culture to be posited in ontological terms, as a given order. Marx's concept of commodity fetishism is predicated upon such purification, paralleling its role within historical materialism more widely. Thus, while Marx posits the emergence of society from nature as a historical emergence, inaugurating 'history' proper, at the same time the society–nature distinction itself is grasped a-historically, for there is no account of its production within the historical process. How can something be deemed to have emerged from something else without any means of distinction between the two being proffered? Only if this distinction is deemed self-evident in the 'objective' characteristics of the phenomena. The distinction is the epistemological ground of Marx's strategy for making commodity relations anthropologically 'strange' and is thus shielded from such strangeness itself. In other words, though Marx historicizes the nature–society relationship, he does not do so symmetrically; 'nature' is historicized by being subject to labour/culture/the social, which modifies/constructs it in historically specific ways. The corollary is that the social itself is not historicized, it is the basis of history rather than its product; while the form of the social may change historically, producing corresponding changes in the form of nature (modes of production), yet sociality itself, the concept of the social, is not viewed as a historical product but as a given; its historical conditions of existence are not explored.[32] If the society–nature divide must be seen as a product of the historical process and not as prior to it, then 'nature' cannot be historicized by being theorized as the construction of a 'society' somehow immune from the consequences of this very move.

In this sense, Marx takes modern humanist ontology for granted; it is deeply rooted in his philosophical anthropology. For all his critical acuity on other questions, in this matter he is an occidental thinker at home in his world; he merely accepts the reality presented to him. From the point of view of a symmetrical account of the ontological organization of modernity, this constitutes a significant weakness of Marxism, which looks much closer to a classically modernist discourse than a critique of modernity in this respect. No doubt this is unfair to Marx in a sense, for it is surely a matter of what can and cannot be thought in a given period, as determined by historical conditions, rather than a failing which might reasonably have been avoided. Nevertheless, both Marx's theory of commodity fetishism and his conception of labour depend upon a *prior* distinction between the natural and the social, which is in turn used to uphold the sovereignty of intentional human agency (or praxis). Marx cannot explain the separation of the domain of the social from the natural without relying upon categories already structured by this separation; he therefore cannot explain how entities and quasi-objects are assigned to one or the other and thereby categorized as subjects or objects in accordance with the dualist ontology. Only 'labour' provides anything like an explanation, but as I have argued, 'labour' itself presupposes the very separation of the social from the natural which it purports to explain, and it roots this separation not immanently within the historical process, but in an external, ontological distinction between 'social' humans and 'natural' animals. In the end, therefore, the asymmetry at the heart of Marx's dialectic ensures that his humanism is not in fact unified with his naturalism but subsumes it.

Cartesian subjects and objects: the unity of phenomenology and realism

The role of humanist discourse in structuring Marx's central category of labour and thus his society–nature dialectic, hence historical-materialism as a whole, has had profound consequences for the subsequent development of social thought. I cannot go into a detailed or exhaustive analysis of these consequences here, but the main point is that Marx's vision of 'society' or the 'social' was rendered internally unstable and contradictory as a result. The category of labour upon which his 'society' depended, and which prevented it from dissolving into 'nature', concealed a metaphysical humanist kernel incompatible with the materialist premises of Marx's method; in short, the dialectic was compromised, and compromised in 'society' itself. This meant that the dialectical conception of 'society' was condemned to oscillate between its two internal poles, humanism and naturalism, each of which should have been cancelled out by Marx's materialism but were instead left partly in place, crossed out but not deleted by an asymmetrical dialectic with a humanist heart. With time these unstable poles were bound to crystallize into irreconcilably opposed, incommensurable positions; thus the social 'subject' and 'object' became at once splintered from each other and dislocated from the human relationship to nature from whence they had arisen. Much of the subsequent development of social theory can be understood as the intellectual shockwave of this fracturing, consisting of various more or less unsuccessful attempts to put Marx's dialectic back together again. Durkheim and Weber, who did more than any other writers to establish sociology as a discipline, illustrate this very clearly. Of course neither of these writers can be summed up in just a few lines, and I do not pretend to have discussed them in anything like comprehensive detail here, but in terms of their core methodological prescriptions and the positions these involve on the nature–society/human–nonhuman nexus, it is clear that they occupy almost antithetical locations on a dialectical continuum defined above all by Marx:

> By emphasising the understandability of human conduct, as opposed to the mere causal explanation of 'social facts' as in natural science, Weber draws the line between his interpretative sociology and the 'physique sociale' [social physics'] in the tradition of Condorcet, which Comte called *sociologie* and Durkheim worked out in such an eminent manner.
>
> (Gerth and Wright Mills, 1991, p. 57)

If Weber is methodologically humanist, paving the way for interpretivist sociology, then Durkheim is systematically realist or 'naturalist' in the positivist sense, advocating the imitation of the method ascribed to the natural sciences. Both, however, rehearse the same basic dualism, though they do so by occupying its opposing sides. At the centre of Durkheim's sociology is the concept of 'social facts'. It is the existence of 'social facts', above all, which for Durkheim justifies the existence of sociology, for the reality of social facts provides the discipline with its *raison d'être* and its object of knowledge. 'Treat social facts as things,' Durkheim famously declared in his *The Rules of Sociological Method*. This was an affirmation that social phenomena have the same ontological facticity as natural phenomena and are therefore knowable in the same way and according to the same method. In its commitment to the notion that there is one valid scientific method for all knowledge and that this is the unified model of a homogeneous natural science, Durkheim's injunction is archetypical of scientism. Indeed, for Durkheim, as Anthony Giddens has noted:

> To regard social facts as things is to perform the act of detachment necessary to recognize that society has an objective existence, independent of any particular one of us; hence it can be studied by methods of objective observation. The most important feature of a 'thing' is that it is not plastic to the will: a chair moves if pushed, but its resistance demonstrates that it exists externally to whoever is doing the pushing. The same is true of social facts, even if these are not visible in the way that a physical object like a chair is.
>
> (Giddens, 1978, p. 35)

> Social facts are external to individuals and exert constraint over them. In *The Rules*, pursuing the theme of the naturalistic character of sociology, Durkheim suggested that these characteristics of social facts closely parallel the externality of the physical world and the constraints or resistances which it offers to our actions.
>
> (ibid.)

In this way, Durkheim's social ontology is built upon a thoroughly Cartesian dualism of mind and matter, for it relies upon a notion of the contemplative individual as a (immaterial) being of consciousness over and against the world, a consciousness which confronts matter as an alien externality and an obstacle to the will. This is by no means an existential given, but on the contrary, it is a characteristically modern sensibility, bound up with the dominance of scientific ways of knowing, which can be contrasted by way of illustration with the sense of the embodied individual as an active being-in-the-world, both inconceivable without matter and inseparable from it. Moreover, Durkheim's position conflates social *collectivity* with the concept of '*externality*' born of sceptical philosophy's mind/body problem. It treats whatever is over and above the individual as 'objective', in the sense of it being like a physical object which confronts the individual 'mind' or consciousness. By beginning in this way from the point of view of a pre-social abstract subjectivity, Durkheim solidifies sociality into externality and process into object; in short, he reifies the social. Hence his underlying ontological individualism leads to the mechanistic objectivism which characterizes his sociological method. This makes Durkheim the theorist *par excellence* of 'society' as a 'thing', that is as an objective unit, totality or system. It is the objective reality of this macro-object, 'society', which, for him, justifies sociology and underpins its methodological autonomy and its scientific character. The unique purpose of sociology, for Durkheim, is to study 'social facts'; it is this which separates sociology decisively from psychology and psychological explanations. That is why Durkheim insists that sociology is a science *sui generis*, or as he puts it:

> A science can be established only when it has for its subject matter facts *sui generis*, facts that are different from those of the other sciences. If society did not produce phenomena which are different from those observable in the other realms of nature, sociology would be without a field of its own. Its existence can be justified only if there are realities which deserve to be called social and which are not simply aspects of another order of things.
>
> (Durkheim, 1964, p. 363)

Exactly so, but if sociology depends so fundamentally upon the coherence of the idea of an ontological domain of 'the social', then it can hardly be relied upon to be disinterested in the establishment of this domain. One must therefore ask critically what role sociology has had (as a set of epistemic technologies) in bringing its object (the transcendental 'social') into being. As Mike Michael has remarked: 'What counted as "natural" or biological, and what counted as social or "human", was influenced by the exigencies of constructing and maintaining institutional and disciplinary boundaries' (1996, pp. 137–8). Indeed, it can no longer be confidently assumed that there *are* phenomena that are specifically and exclusively social. 'Social' phenomena are always already technological, political, economic, legal, scientific and indeed natural. The attempt to specify some essence within these constitutively interwoven phenomena which is specifically 'social', or to designate a level at which all these things are resolved into a general sociality or relationality which encompasses them in some totality, or which is relatively autonomous of them, is to create a pure abstraction, referring to nothing that actually exists in the heterogeneous world with its pervasive hybridity. The purpose of such purification is self-affirmation; to create the object which justifies the existence of social science itself. Such affirmation is not the exclusive preserve of sociology; the discipline has merely been complicit in the epochal practice of meticulously separating the 'social' and the 'natural', rendering their mutuality in any actual case invisible to modern thought. Durkheim did more than anyone to ensure that sociology developed in this direction.

Not only does Durkheim reify 'society', but he also treats 'it' like a natural organism, with 'needs' that 'it' can somehow induce individuals to fulfil blindly even as they pursue their own individual purposes (i.e. functional explanation). In this sense Durkheim posits a form of *social determinism*, in which the intentions and understandings of social actors are sociologically irrelevant. This social determinism laid the conceptual foundations which were later rehabilitated by semiotics/structuralism, albeit with a synchronic culture-as-language now taking the place of 'society' as the sovereign abstraction in which all social explanation must ultimately be grounded. Indeed, though Durkheim has been very widely criticized for his neglect of human 'agency' by a sociological tradition that has often preferred to be identified more closely with Weber, the structure of Durkheim's thought has been massively influential in shaping subsequent notions of 'society'; as Patrick Joyce has argued:

> The tendency to think of social structures, relationships, and processes as abstractions which yet become in one sense or another real entities, relatively autonomous of other sorts of human activity, and systematic in their operation, has for some time been in question, yet it is still strikingly persistent, in many subtle ways.
>
> (2002, p. 1)

> The assumptions of a thing-like social are so embedded that the (still radical) argument that the epistemological institution of abstract space may more productively be seen as the cause and not the result of these supposedly social phenomena is not entertained. So this old, solid, ontological social still seems to be around.
>
> (ibid.)

Though this is not the place to go into a catalogue of evidence to substantiate the point, it should be fairly obvious that sociology, and 'macro' sociology in particular, have tended to reproduce this ontological conception of the social rather uncritically in their approach. In this respect, Marx again stands apart in offering a more careful position than much of what has followed him; his critical view of the apparent externality of the social as a historically specific consequence of a particular mode of organization of production can be contrasted sharply with Durkheim's elevation of this same reification into an ontological and methodological principle. Whereas Durkheim's (scientific) naturalism therefore is clearly a species of positivism, Marx's naturalism is historical and dialectical; he problematizes precisely what Durkheim institutionalizes. That is why Marx rejects causal and functional explanations of social phenomena, whereas Durkheim regards these as precisely the forms of explanation which distinguish sociology as a science.

In his commitment to the idea that social phenomena are epistemologically equivalent to natural phenomena, Durkheim's sociology could not be more sharply contrasted with that of Weber. At the core of the Weberian critique of Durkheim is the charge that, by treating social phenomena as though they were 'things', as though they were like the objects of natural science, or at least similar enough to justify a realist-positivist epistemology, Durkheim neglects the 'cultural' nature of social life; that is, the meaningful character of social actions and the reflexivity and subjectivity of social actors. It is precisely this meaningful aspect of social life that Weber's sociology emphasizes;

indeed, this is the basic enabling assumption of all post-Weberian interpretative/phenomenological sociology – that social phenomena are incommensurable with natural phenomena. This is to have come full circle, to the point at which Weber, in his antithesis to Durkheim, ends up performing the same onto-epistemological dualism that animates Durkheimian naturalism. These apparent opposites are united not merely in their dualist approach to nature and society, but in their essentially ontological approach to this dualism, for neither acknowledges that the distinction is itself historically constituted, that it does not follow self-evidently from the nature of things – for this perception is also historical, perpetually reconstituted by a specific ontological politics. Hence Weber's 'interpretative sociology' or 'sociology of understanding' was, for him, 'a unique approach of the cultural sciences, which deals with man rather than with other animals or with lifeless nature'.[33] The assumption is that human conduct alone is 'meaningful' and 'understandable'; thus, whereas the objects of natural science must be studied according to the method of causal explanation of facts, the subjects of social/cultural science must be grasped according to the method of interpretative understanding, that is, in terms of the intentions and understandings of the subjects involved. This echoes, at least in part, the Enlightenment view of 'man', in which the reasoning and rational individual is the basic unit of society. The putative self-understanding of such individuals provides the terms in which interpretivist sociological explanation is ultimately conceived. It therefore enthrones the modern 'subject' as the constitutive unit of an intersubjective 'social'. In Weber, then, the dualism of nature–culture is embedded at the level of his philosophical methodology, in how he distinguishes the 'sociology of meaning' from the natural sciences. Hence the whole methodological orientation of Weber's sociology is predicated upon a humanist ontology, the notion that there are two incommensurable realms: one of human subjectivity, meaning-for-itself, 'culture', the other of nonhuman objectivity, matter-in-itself, 'nature'.

Of course, Weber is a complex sociologist, in many ways far more complex than Durkheim, and the boiling down of his methodological position to this humanist ontology is far from simple and straightforward. As Giddens points out:

> The genealogy of Weber's methodological essays is complex, and they must be placed within the framework of the then current controversy over the relationship between the natural and the 'human' or social sciences. Whereas Durkheim was steeped in a tradition of positivism which reached back to well before Comte, no directly comparable tradition existed in German social thought. The lengthy and complicated debate which arose in Germany over the status of the sciences of man thus explored issues which remained largely quiescent in French history and social philosophy.
>
> (1971, p. 133)

Nevertheless, in the end, Weber's immersion in this debate led him to the essentially Kantian conclusion that 'The social sciences are necessarily concerned with "spiritual" or "ideal" phenomena, which are peculiarly human characteristics which do not exist in the subject-matter treated by the natural sciences', (ibid., p. 134) Just as Marx's humanism is most visible in his resort to metaphysics in order to legitimize a spurious distinction between human and animal labour, Weber's humanism is most apparent in his not dissimilar distinction between 'social action' and 'mere behaviour':

> Social action, in the Weberian sense, is distinguished from mere behaviour. Action which is oriented towards inanimate objects, for example, does not qualify. Action is 'only social if, and then only in so far as, it takes account of the behaviour of someone else'. Further, action involves more than mere contact of individuals. Being jostled about by people in a crowded department store does not, according to Weber, constitute social action. Action of a social character 'is confined to cases where the actor's behaviour is meaningfully oriented to that of others'. In short, only behaviour which is intentional, and which is oriented toward the behaviour of other human beings, qualifies as social behaviour.
>
> (Tucker, 1991, p. 45)

This underlines the extent to which Weberian sociology is predicated upon an ontology/epistemology which systematically excludes nonhumans from consideration. With breathtaking circularity, 'sociality' itself is deployed as the mechanism of this purification, for whatever is not inter-subjective, in the sense of existing in the world of meanings between human beings, is simply not 'social', by tautological definition. By this sleight of hand, Weberian sociology conjures up a purified dualist world inhabited only by human beings and inanimate objects. This raises the question – why is behaviour which is oriented towards nonhumans not 'social'? Indeed, how does one even hope to determine what, if anything, a given instance of behaviour is oriented towards? And where does this dichotomy leave such knowledge formations as contemporary cultural primatology, which employs a broadly interpretative – even 'ethnographic' – approach, albeit towards the behaviour (action? praxis?) of nonhuman animals? These animal behaviours are not the inert objects of natural science, ostensibly amenable to strictly causal explanation, but nor is it possible to rely on the linguistically mediated intentions of the actors in order to understand their action. Indeed, when one problematizes modern (occidental) ontological presuppositions about what makes up our reality, it is difficult to avoid the conclusion that human practice manifests an inextricable 'mangle' of multiple and overlapping orientations, influences, intentions, unintended consequences and relations.[34] These are oriented not just towards human subject-agents or nonhuman object-natures, but to individuals, groups, collectivities, institutions, a multiplicity of distinct objects, artefacts and technologies, a vast range of different kinds of nonhumans, animals, plants and quasi-objects, to the material and corporeal characteristics, 'textures', qualities and capacities of those nonhumans, to their economic values, to their various meanings, to technologies and the material relations of technologies, to embodied practices and routines, to a variety of different systems of beliefs, to distinct spatial formations, and so on *ad infinitum*. Not all or even most of these relations are rational, coherent or intentional. Nor is there any reason to think that this heterogeneous anthropological matrix of objects, quasi-objects, subjects, quasi-subjects, in short, hybrids of all descriptions, can be reduced, ultimately, to a fundamental distinction preceding all others between just two types of things – subjects and objects; it is absurd. Yet this is exactly the proposition that any absolute ontological distinction between the natural and the social – or, which is the same thing, any absolute epistemological distinction between the natural and social sciences – is founded upon. When this way of organizing the world is recognized as a contingent discursive formation rather than an unquestionable given, then it can begin to be understood genealogically, in historical and political terms.

Weber and Durkheim then, at the most fundamental level, are far more alike than distinct; the humanism and scientism respectively which characterize their work turn out to be different faces of the same ontology. In the present intellectual climate, with its obsessive pursuit of the 'new', it is not unthinkable that a question might be raised as to the relevance of any critique of Weber and Durkheim for contemporary social thought. In response, I should emphasize that this is not just a matter of intellectual history, for in the shape of constructivism or phenomenology on the one hand and realism or scientific-naturalism on the other (both being understood in the broadest sense), this dualism continues to define a great deal, indeed most, of contemporary social thought. For such thought, nature is still an absent other, it is what happens somewhere outside of the semiotic/meaningful/cultural/social, and can for that reason be either left to the natural sciences or theorized in 'cultural' terms, but never grasped symmetrically in a vision of a collective that is as natural as it is social. Marx is genuinely different, offering a dialectical model (albeit an asymmetrical one) of the nature–society nexus, and for this reason he must still be a key point of reference for theories of nature–society relations.[35] On this note I want to make a chronological leap to actor network theory, which, though often associated with 'micro' sociologists instinctively unsympathetic towards the Marxist tradition, can productively be read as a powerful rejoinder to Marx's dialectical conception of the nature–society connection.

The reconnection of culture and nature: actor network theory (ANT)

> There are no humans in the world. Or rather, humans are fabricated – in language, through discursive formations, in their various liaisons with

> technological and natural actors, across networks that are heterogeneously comprised of humans and nonhumans who are themselves so comprised. Instead of humans and nonhumans we are beginning to think of flows, movements, arrangements, relations. It is through such dynamics that the human (and the nonhuman) emerges.
>
> (Michael, 2000, p. 1)

Taken together with key insights emerging from the field of science and technology studies (STS), with which it has been closely associated, the project of actor network theory (ANT) broadly conceived has underpinned an increasingly widespread attempt to study modernity in a consistently symmetrical manner. This has meant rejecting the usual privilege granted to western science's self-conception of the uniqueness of its particular way of knowing, which claims to afford direct and unmediated knowledge of nature, and which has traditionally for that reason been exempted from properly anthropological treatment.[36] Sociologists, anthropologists and others pursuing this agenda have been at pains to show how the production of scientific knowledge remains at all times *inside* what social science habitually thinks of as 'culture' or 'society'. Hence it is not unusual now to encounter the argument that *natural* facts too are socially-historically constituted, more specifically within and not outwith the local knowledge-practices of science. These facts, though unalterably local in their production, are able to become seemingly universal by the progressive (micro-political) extension of the networks of practices and technologies wherein their 'truth' is demonstrated – by the proliferation of the relevant theatres of proof.[37] This proliferation is of epochal significance well beyond the laboratory; it is the techno-political condition of possibility of 'modernity'.

I broadly accept this analysis, but it seems to me that it does not go far enough, for it responds only to the least problematic side of the implications of Latour's position, that is, the side broadly compatible with an extension of the essential principles of constructivism and 'cultural' thought. For if natural 'facts' are 'socially constituted', then we must also acknowledge that social 'relations' and 'contexts' are shot through with natural 'facts'. That is, we must challenge the tendency of the logic of social construction to assume that matter has no 'agency' (that matter is 'object' and not 'subject') and that 'contexts' and 'relations' are necessarily 'social' or inter-subjective. It then begins to become meaningless to argue that natural facts are 'socially constructed' because we can no longer invoke the nature/culture, subject/object dualism which at bottom renders this statement meaningful. Faced with this dilemma, some have tried to sustain the constructivist agenda by universalizing it, asserting simply that we live in a radically constructed world; but by contrast with what other does this constructedness gain its identity? Without invoking some non-constructed other, this claim cannot be differentiated from the apparently opposite claim that we live in a radically given and natural word. It is not sufficient then to theorize the hybridity of the world as it were 'positively', in an ontological mode, for when symmetry is pursued consistently, this mode of explanation soon collapses into tautology. The theoretical identification of hybridity is merely a strategic starting point, from which it is necessary to turn genealogically to the processes of purification by which hybrid being is fashioned into an asymmetric and ordered world.

Science studies' commitment to reveal the political dimension of techno-scientific knowledge and practice, then, from the point of view of a rigorously symmetrical approach, can at times look rather like a covert extension of the domain of 'culture' leaving the semiotic logic of 'culture' largely unchanged. As Christopher Pinney has insisted, what is required is something at once more destructive and more radical, questioning the very foundations of the 'human sciences'.[38] No doubt there are several strategic routes by which such a thing might be attempted, but I want to focus upon one in particular, namely a turn to nonhuman animals, or a 'zoological turn'.[39] I contend that nonhuman animals are a highly liminal category possessed of a capacity to disrupt the coherence of the dualist structure of humanist ontology; indeed, they are a unique kind of other, which poses a special challenge for humanist discourse. Animals, in other words, have not been comprehensively rationalized by modernity; they have partially eluded the logic of its pervasive dualism and have persisted as a living contradiction, radical alterity or liminal case, a 'third estate' tendentially transgressing the boundaries and exposing the contingencies of the humanist mode of order. As Mike Michael puts it:

> As we have seen, technological actors can be traced back to the activities of human actors who are themselves the partial products of the pre-/pro-inscriptions of technological nonhumans (and so on *ad infinitum*). In the context of encounters between human and nonhuman entities, it is possible to attribute, albeit circumspectly, agency to the latter by virtue of the fact that they have been partly constructed by humans. As such, the wills, desires and goals of those humans can be said to reside in, to be embodied in, those technological nonhumans as they go about their routines of disciplining human bodies. Indeed, as Latour argues, technological artefacts are the most moral and moralising of characters, structuring human action/behaviours in all sorts of ways. In contrast, it is, at this point in history, considerably more difficult to do the same for 'natural' nonhumans like animals and local environments. In a sense, the status of 'natural' nonhumans is more opaque in our current Western networks (though historically the 'agential' status of the 'natural' has, in the past, been more readily accepted).
>
> (1996, p. 131)

To my mind this is absolutely critical. It is an acknowledgement not merely that the ontological status of 'natural nonhumans' is uniquely problematic for modernity, but that it remains problematic for actor network theory, despite its attempts to provide a theoretical means to overcome human–nonhuman dualism. Michael hints that this is because ANT has tended to privilege technological nonhumans in its accounts of human–nonhuman relations. But the problem is not just that such 'constructed' nonhumans have been favoured empirically by ANT as objects of research interest, but that they have become the defining model or archetype for ANT's understanding of nonhumans in general, as is clear from its favouring of the term 'socio-technical relations'.[40] Indeed, it sometimes seems as though ANT consists of little more than a thorough reckoning of scientific knowledge-practices and technological assemblages into an otherwise little altered conception of 'the social'. One might well ask how far this really takes us beyond Marx, who also recognized that 'things' as bearers of social relations could take on a seemingly independent power, a 'second nature' over and above human agency.[41] It cannot be assumed that such accounts are adequate for understanding the positioning of other kinds of nonhumans, as this would involve reifying the category of the 'nonhuman' by effectively treating its occupants as though they were homogeneous. Such reification is tantamount to inadvertently re-inscribing humanist discourse, which regards all nonhumans as equivalent in their nonhumanity. Michael himself does not avoid this entirely even in the foregoing passage, for his suggestion that both local environments and nonhuman animals as ' "natural" nonhumans' pose the same kind of challenge to 'our current Western networks' is questionable, as is the category of ' "natural" nonhumans' itself, for it has to invoke the nature/culture divide even in its attempts to move beyond it.

This kind of problem has no doubt been one of the reasons behind the attempt to move beyond actor network theory by many of those sympathetic to its agenda, who have sought to develop a 'post-ANT' or 'after-networks' approach.[42] This book draws upon some of the insights of such work but represents a slightly different response, one which proceeds by historicizing the problem and mapping its conditions of possibility, rather than trying to solve it in an ontological mode by means of further theoretical innovation. It also makes the strategic decision to identify the (species) relations between human and nonhuman animals as the absent, unstable and vulnerable centre of modern culture/nature dualism. Thus, rather than attempting to challenge this dualism by demonstrating the 'agency' of nonhumans and their complex roles as 'actors', thereby inadvertently reinforcing what Daniel Miller refers to as 'the tyranny of the subject',[43] my objective is to trace the work of humanist discourse and practice in (re)constituting the domain of the 'cultural' and hence the 'subject' in contradistinction to a nonhuman 'nature' – to trace these processes of purification from the critical perspective of a generalized symmetry.

Though totalizing in its logic, the discourse of 'culture' is never an effective totality, but must perpetually reproduce the conditions of its own cogency. With this in mind, it makes critical sense to analyse the means by which marginal entities, transgressive experiences and hybrid encounters have been (partially) managed and contained by disciplinary technologies within the ontological project of modernity. In animals we recognize an other-worldliness which is nevertheless a kind of 'subject' like ourselves; while so far from being 'human', animals are in many ways so close to being 'human' that they seem to pose an existential question not merely as to the boundaries of 'humanity', but also to our deeply held notions of its value.[44] How is it possible to conceive of the world in terms of 'humans' and 'nonhumans', or even in terms of 'societies–natures', when humans are a species of animal among countless species? What is the ontological politics of defining 'non-humans' against a single species, when nonhumans include both apes and satellites – are satellites really more like apes than apes are like humans? The existence of nonhuman animals enables us to ask these deeply anti-humanist questions, and if only for that reason animals problematize anthropocentrism in a far more profound and radically symmetrical way than do technological objects.

Far from being subsumed under a theory of objects, things or technologies, then, animals need to be grasped in their specificity and viewed as pivotal in the humanist project, that is, in defining the boundaries of the human for modernity. Their importance is bound up with the fact that human relations with other species are not mediated primarily by linguistic communication, which militates against a residually semiotic (hence constructivist and humanist) approach to theorizing these kinds of nonhumans, nudging us instead towards something like a notion of 'inter-corporeality'. This is a materialist grounds for conceiving of relations between beings. Hence, as Mike Michael puts it, drawing upon Donna Haraway:

> To engage in those non-linguistic processes of communication with animal others is *to diffuse one's self* – to admit of its multiplicity and dispersion . . . The identity that is thus generated is one that no longer follows the linearities of narrative, but is one mediated by and realized in the many channels of human–animal communication . . . it is the form of non-linguistic communication that becomes the content of human identity in the association between human and animal companions. So, animals, in all their other-worldliness, enrol humans; but the identity that they offer them is of a different order from the typical models of the human social self.
>
> (1996, p. 144)

For this reason, our encounters and relations with nonhuman animal others have the potential to induce destabilizing and transformative reflections upon our own 'nature' as humans in a way that is very different from our encounters with technology. It is not 'cyborg' technologies then which are at the frontier of our existential encounter with a possible post-human other, but 'beasts', 'mutants' and 'monsters'. Thus the typical focus upon technological nonhumans needs to be displaced by emphasizing the critical importance of living nonhumans, which may well have been modified by human agency but are not in any primary sense human 'constructs'. Writing living nonhumans into modern ontology, or rather, tracing the politico-ontological processes through which they are policed and managed by being perpetually made absent and invisible as 'nature', is a significant departure from what has been the main theoretical and empirical focus of ANT.

Have I constructed a straw man to suit my argument, however, or is there really in ANT-influenced work an overemphasis of technological nonhumans at the expense of nonhuman animals? I think it reasonable to maintain that there is, at least tendentially. Like Marx, ANT defines 'nonhumanity' overwhelmingly in terms of produced objects. There are remarkably few exceptions, perhaps the most notable being Michel Callon's widely discussed analysis of the co-constitutive agency of the scallops, researchers and fishing community in St Brieuc Bay.[45] Another is Ashmore's playful exploration of the possibility of a symmetrical analysis of the relations obtaining between a cat, a human, and a catflap.[46] One of the more recent is John Law and Annemarie Mol's interesting discussion of the enactment of Cumbrian sheep as 'actors'.[47] Even in such instances, however, where 'natural' nonhumans have indeed been theorized, they have not been theorized *historically* and genealogically. The theoretical point has skilfully been made that animals are intrinsic to the heterogeneous networks which constitute our 'material-semiotic' world, but this has not been identified for what it should be, a mere starting point for a genealogy which explores in terms of material and ontological politics how the radical commensurability and interconstitution of human and nonhuman forms of life are perpetually deleted, contained and disciplined as a condition of our 'culture'.

Equally, ANT concepts developed predominantly in relation to technological nonhumans have often been extended with little modification to other kinds of nonhumans, a strategy which seriously underestimates and obscures the specificity of living nonhumans. For example, ANT's preoccupation with the problematic of 'agency' and its distribution between humans and (technological) nonhumans has tended to stifle any thoroughgoing consideration of how meaningful the concept of 'agency' is when stripped of its anthropocentric architecture. As a result, 'agency' is often extended within ANT beyond the conceptual framework which gives it coherence. Indeed, if 'agency', as I suggest, is a discursive product of human/nonhuman dualism rather than somehow ontologically prior to this divide, then there is something deeply problematic about the notion of extending the concept of 'agency' to nonhumans, for rendered thus it would no longer *be* 'agency'. Thus to insist upon the agential status of nonhuman animals is effectively to anthropomorphize them, and is therefore no solution. As Michael observes in reference to the field of environmental ethics: 'In this latter case, "nature" is merely "humanized": "nature" is attributed characteristics typical of humans conceived in humanistic terms.'[48] In some cases this problem is cleverly side-stepped via a re-conceptualization of agency as an emergent element in the network, but I think there is a real danger of retaining agency by reinventing it, for this tends to obscure the work that anthropocentric 'agency' has historically performed. Indeed, this sort of redefinition of the terms is one of the main weaknesses of ANT's perhaps overly 'theoretical' approach. It would be better if the role of the concept of 'agency' in the constitution of humans vis-à-vis nonhumans be problematized by being grasped historically.

There is also in ANT an insufficiently central and systematic concern with power. It is not that ANT-influenced theorists have not considered power, for several of them certainly have, and seriously, as the very useful concept of 'ontological politics' demonstrates;[49] but the analysis of power, though intrinsic to the emergence of certain aspects of ANT from Foucauldian roots, has nonetheless been insufficiently central to the development of the perspective. As a result, the concept of power utilized by ANT theorists is not always as consistent with the implications of the approach as might be desired, but sometimes seems imported from elsewhere; it often consists in a salutary emphasis of material, embodied, and especially technological power, but without any systematic attempt to situate the analysis of such power within a longer-wave sense of historical changes in the rationalities and modalities of power. A sense of this long-wave historicity of power is present, albeit only abstractly, in Latour's *We Have Never Been Modern*, but is frequently absent or underdeveloped in later work within the ANT paradigm. Power might well be an effect of the mobilization and stabilization of large, complex and heterogeneous networks, but the usefulness of such formulations is very limited if posited in a historical vacuum.

Moreover, it seems unwarranted to view power chiefly as an effect or emergent property of networks rather than the means of their (re)production. Power, after all, is productive, and particularly modern disciplinary power. It is by no means clear that this form of power observes any ontological distinction between human and nonhuman in what Foucault refers to as its 'art of distributions'.[50] Rather it produces this distinction through its own techniques, which arise in the practice of various disciplinary institutions before being diffused; hence such techniques must themselves be seen as generative of the principles of ontological differentiation, of subject and object, culture and nature, human and nonhuman. There is clearly no case for privileging *either* networks *or* power in some causal hierarchy; the key thing is to recognize their simultaneity, their mutual constitution, in practical situations. The networks are themselves political technologies, they are the machinery of power, but power in turn produces the networks.

Such problems have been mitigated in some cases by the integration of the Foucauldian problematic of 'governmentality' within a broadly ANT approach or vice versa. A focus upon governmentality is essentially a genealogy of power; it directs attention not towards relations of force between what are taken to be pre-existing, given, known social objects and knowing subjects, but instead towards how these configurations of objects and subjects of power-knowledge are brought into being through multiple, overlapping, contingent processes.[51] Thus governmentality takes as its object of enquiry not power as such, viewed as a brute force of coercion between separate sovereign entities, but rather the conditions of possibility of particular modalities of power understood as organized around historically specific rationalities.[52] This is close to an ANT view of power, except that the latter attempts to explain relations of power in terms of multiple enactments of ontology, whereas the study of governmentality critically reconstructs the rationalities of power in terms which are more consistently epistemological. Thus the analysis of governmentality is concerned with historical regimes of knowledge, or as Mary Poovey puts it, with

> the epistemological field that allows for the production of what counts as knowledge at any given moment. This field changes over time – it has a history – and has its own distinctive categories – categories like domains, genres, discourses, disciplines, and specific rationalities.
>
> (1995, p. 3)

This has led to some incisive analyses of the mutual constitution of formations of knowledge and material power, political rationalities and ontological discourses, tactics and technologies.[53] The empirical range of the work undertaken within this approach is still fairly limited, however, and so far it has manifested a marked lack of attention to the politics of 'natural' nonhumans, that is, to the disciplinary technologies by which living nonhumans are governed and how these articulate with, underpin and condition the political-ontological government of humans in historically specific ways.

One of the great strengths of the dialectical approach was that, properly understood, it insisted upon the game of actuality and possibility, tracing the immanent contradictions within material and social formations in order to identify optimal moments of critical engagement and transformative potential: the point, as Marx famously stated, was to change it. For all its faults, the power of the concept of the 'capitalist mode of production' is still palpable in this respect, serving as a reminder of the strategic and political potential of 'macro' historical abstractions. Indeed, 'macro' concepts were always in part intended as strategic abstractions; it was not really a question of whether things corresponding to these abstractions existed in the world, but of what we could do with these abstractions; this is the reflexive lesson of 'performativity'. With this in mind, having challenged the autonomy of the human/social/cultural and grasped the inseparability of the 'social' and 'natural', can we make this the basis for a post-humanist form of knowledge with as much political coherence as what we used to call 'macro-social' thinking, and can we do so without a return to humanist and reified conceptions of social relations? It seems to me that this has to be done, if the decentring of the social is not to be bought at the cost of the final dislocation of social analysis from any transformative collectivist politics. While the deconstruction of reified abstractions is necessary, indeed essential, deconstruction without end and for its own sake eventually risks sanctioning an individualist politics of despair. Ultimately therefore the challenge has to be to reconstruct 'sociology' as such, but to do so without returning to the humanist dualisms of neo-Weberian/Durkheimian models of 'society', or indeed 'culture': what is required is nothing less than a new kind of sociology. I do not try to accomplish this here, but I do try to clear the ground for such a project and to point towards one possible way forward.

2 Milk and modernity Part I
Commodities, networks and monopolies

> Milk, of all agricultural products, is the most difficult to market. It is liquid, heavy and awkward to transport. It is also, as 'nature's finest food', a first class medium for bacterial growth and highly perishable. Additionally, as another significant feature, it is produced in family farming units with, usually, either big buyers to contend with or uncertain distant markets to face.
>
> (Empson, 2000, p. 10)

> We trouble ourselves little to think of the difficulties which have to be encountered in supplying so changeable a liquid in a condition to satisfy the fastidious requirements of the public.
>
> (Bond, 1893, p. 32)

From town cows to milk trains

The marketing of liquid milk as a commodity is very much a phenomenon of the late nineteenth and early twentieth centuries, with its origins in the railway expansion of the 1860s and 1870s. Prior to this, the highly perishable nature of the substance meant that it could not be transported any distance to consumers without becoming sour and inedible. The action of micro-organisms within the milk rendered time and distance impassable barriers to its commodification on anything but a local scale. For this reason the industrialization of the late eighteenth and early nineteenth centuries had little or no effect upon milk distribution, and modernity did not begin to penetrate the milk trade until the later part of the nineteenth century. Until this time all milk production was local by necessity; rural dairy farmers had little choice but to sell their milk to the immediately surrounding areas, while the large towns and cities obtained most of their milk either from town cowhouses or from the few farms located on the outskirts of the city.[1] In the 1850s, there were about 24,000 such cows in London.[2] Manchester obtained much of its milk from farms located along the Bridgewater Canal, and consequently relatively few cows were kept within the city.[3]

It is the steam engine and the expansion of the railway network which are conventionally thought to have been responsible for transforming this situation, for they began to mitigate the deleterious effects of time and distance upon milk, by effectively shrinking distance in relation to time. The length of line open for traffic doubled between 1854 and 1873 from just over 8,000 miles to just over 16,000 and still further to almost 21,000 miles by 1893.[4] This was not a simple causal relationship in which the development of the railways determined the growth of milk marketing in some unilinear process, but rail transportation was certainly a crucial historical condition of possibility of the modern milk trade. One contemporary commentator put it more strongly:

> The milk trade is very modern indeed, so far in any case, as its wholesale features and the transmission of milk by rail is concerned. It did not and could not come into existence prior to the advent of the railroads. It is, in fact, essentially an offspring and consequence of the current method of transit by means of steam – a marvellous instance of the triumph of mind over matter.
>
> (Sheldon, 1893, p. 28)

Other critical factors included the trend towards urbanization, which meant that the demand for milk in the cities could no longer be adequately met by town cowhouses and peripheral farms. This was exacerbated by the outbreak of rinderpest during the years 1865–67, which wiped out a large proportion of the town herds. As a drastic measure to control the further spread of contagion and as part of the Cattle Diseases Prevention Act, the government ordered the compulsory slaughter of all of London's remaining urban cattle, an act which was soon repeated in several other large cities.[5] Hence the major cities suddenly found themselves desperately short of milk, and within the space of a few years the entire economic geography of the milk supply began to be transformed as a result, with town cowkeepers reinventing themselves as retailers and obtaining their milk from the country by rail. At the same time, the agricultural crisis of the 1870s precipitated to a significant extent by the emergence of significant competition from foreign imports, was leading many farmers towards conversion from arable to dairy farming, and especially to the sale of liquid milk, which was widely regarded as a more secure branch of farming due to its relative insulation from the effects of foreign competition.[6] Thus the conditions were in place for a dramatic expansion of the railway milk trade, as the swelling urban population provided a new and growing market for increasing numbers of rural producers.

In January 1865, the Great Western railway carried just under 9,000 gallons of milk; by January 1866, this had increased to 144,000 gallons, and by 1900 this railway alone was carrying around 25 million gallons a year.[7] This exponential increase in the railway milk traffic alongside the sudden demise of the town herds created a whole new set of problems and opportunities, technical, logistical, and sanitary. Indeed, before milk could be supplied to the urban markets on a mass scale and its transformation into a modern commodity completed, a complex knot of interconnected problems had to be managed or overcome. Before it could become an exchange-value on an industrial scale, milk had to be made more durable in relation to time, or if this was not possible, then the action of time upon milk had to be somehow frozen. Exchange-value is a formal and synchronic abstraction inimical to the recognition of natural time, the reversible and particular time of substance, ineluctable material change, transformation and decay.[8] For commodity markets to function, this temporal modality must be carefully managed and regulated; where possible it is excluded, at least formally, and replaced by a manufactured, irreversible time designed around the rhythms of capital accumulation.[9] Where the material action of time is too slow, for instance, then techniques such as built-in obsolescence ensure that the necessary rhythms of consumption are maintained; where the action of time is too rapid, then it must be delayed and suppressed by various means. Such techniques for the management of time are integral to modernity's codification of time as 'progress', as linear and rational time.

In this way, the time of progress and the time of exchange-value or capitalist production are intimately connected; both involve the projection of a humanist teleology, the fiction of a purely social time, progressing somehow separately from and independently of reversible time, recognition of this latter being pushed – like concrete labour itself – to the margins of the social imagination. Thus the transformation of the milk trade into a modern system of commodity exchange required that the reversible time contained within milk be rendered innocuous, and that the potential for material change invested within the substance be somehow foiled. At first this led to the use of various chemical preservatives which it was hoped would prevent the milk from going off during transportation, and then subsequently to numerous attempts to develop a suitable form of refrigeration technology.

The adoption of the capillary refrigerator was a milestone in this respect. Recommended by the judges at the Cardiff Royal Show in 1872, a similar design had previously been used by brewers for cooling beer.[10] Ten years later it was in general use, allowing milk to be sent distances of around two hundred miles. Great emphasis was placed on the necessity of cooling milk immediately after it was drawn from the cow in order to delay the action of bacteria which would otherwise begin to multiply within the milk soon after its exposure to the air. The demand for 'milk warm from the cow' which had served in the popular wisdom as a rough indication of the apparent 'freshness' and authenticity of milk during the years of widespread adulteration by town milk sellers, was now quickly relegated to the small towns and villages that still obtained their milk direct from local farms. In contrast, all possibility of supplying the large urban markets was dependent upon milk which, far from being warm, with all the directness and immediacy of the relation between human and animal that this implied, was instead as cold as possible, its temperature having been reduced to around 40 degrees Fahrenheit by the use of the new cooling technology.[11] The traditional suspicion of human intervention at any stage between the cow and the consumer, which had been thought of as 'meddling' and associated with practices of adulteration and with watering down of the milk, had begun to be displaced by a suspicion of the milk in its unmediated state.

This was not an irreversible or unidirectional historical shift, but it was one twist in a shifting dialectic of trust and anxiety in relation to the human and the nonhuman, which persisted in different modulations throughout the period of the emergence of the modern milk industry, as the particular organic properties of milk and the state of knowledge regarding the impact of temperature upon bacterial life interacted with the geography of milk production, with regional differences in soil and climate, with urbanization and with the steam engine. Each of these diverse entities and networks entered into the historical development of milk marketing; their heterogeneous interrelationships were intrinsic to the commodification of milk and to its transformation into an exchange-value.

Organizing space: technologies of measurement and transportation

The early system of distribution developed on an ad hoc basis in response to the huge increases in the quantity of milk brought by rail into the cities after 1865. As such, it was subject to no overall coordination and very little regulation until the early twentieth century. Typically the dairy farmer despatched the whole of his milk in his own churns to one of the wholesale dealers operating in the city. As soon as it arrived, the wholesaler received it before immediately selling it on to retailers at the station, who took away what they could sell in their own receptacles which they loaded onto carts; from these, milk was sold to the customer directly, speed being of the essence.[12] Usually the majority of the farmers' cans were emptied at the station and could at once be moved to the other platform ready to be loaded onto the next train for return. Despite the relatively limited practical role of the wholesalers, their position as intermediaries was strategically crucial and their economic power grew rapidly and soon exceeded that of the producers, for whom they were the critical gatekeepers providing access to the urban market.

In part, this was due to the political-economic structure of the market; there was very little competition between wholesalers, who agreed the price collectively and were frequently therefore in a position to offer as much as they wished to the farmer and to demand as much as they wanted from retailers. Each farmer usually had access to one wholesaler only and would sell the whole quantity of milk to them, any excess being distributed locally; thus they had little bargaining power when it came to agreeing the price. Competition was further limited by the physical construction of the railway platforms, there being only very limited space, which prevented a proliferation of dealers who might have been more likely to undercut the established wholesalers.[13] Equally, there was only enough space for a limited number of retailers and each dealer had his regular customers, thus stifling the emergence of a market proper.

For these reasons the business remained very substantially in the hands of the wholesalers throughout the last quarter of the nineteenth century, giving rise to a lasting and often bitter conflict not just between producers and dealers, but also between farmers and those other crucial intermediaries, the railway companies. Disputes usually centred upon wholesale prices and rates of carriage, but also encompassed such matters as whose men were responsible for which aspects of the loading and unloading of the milk churns during transit and distribution, and who was liable in the event of any losses due to souring which might occur during transportation.[14] The railway companies were formally obliged by their contracts to assist in loading the churns onto the trains, but this condition was seldom adhered to.[15] Moreover, farmers frequently accused the railway companies and dealers of leaving milk lying around for extended periods on railway platforms and then refusing to accept any blame when the milk was later found to be beyond use. This was particularly problematic in the summer months, when the milk might be exposed to the sun and to higher air temperatures. Wholesalers, on the other hand, claimed that it was farmers who often left milk lying around for long periods after milking, instead of immediately cooling it ready for transportation.

Such disputes were inflamed by successive legislative attempts to crack down on adulteration, culminating in the 1901 Sale of Milk Regulations and the setting of a legal milk standard, under which samples of milk would regularly be taken for testing, either from the railway platform or from the retailer. In cases where the sample fell below the legislative standard of composition for genuine milk, it was frequently unclear which of the many agents involved at the various stages in the passage of the milk between the cow and the retailer was properly responsible. In many cases no doubt the cow itself was responsible, for farmers had long been aware that the constituents of milk could sometimes vary substantially without any apparent human explanation, but the new legal definitions rendered this 'appeal to the cow' illegitimate, for as part of nature the cow was not recognized as an actor in juridico-political terms. Thus it was left to the local Food and Drugs Inspectors to investigate each case forensically until responsibility could be ascribed to the appropriate human agent.

There was little trust between any of the parties, and as yet there were insufficient formal protections to render trust a luxury. It was a common belief among farmers, for example, that railway employees were accustomed to helping themselves to some of the milk while it was in transit, and there was a recurring debate on whether to adopt closed churns rather than continuing to employ the open churns which were then widely in use. The matter was complicated by the belief that it was necessary to let milk 'breathe' if it was not to lose its 'unique taste', which led to debates on the wisdom of attempts to design churns which were ventilated but at the same time closed to possible contamination.[16] The Board of Agriculture was so exercised by this problem that it sent a circular to churn makers advising that the total weight of the churn should be stamped on, that the lid should be made easy to seal, that it should be constructed to prevent the removal of milk or the addition of water while it was sealed, and that the whole container should be dirt- and dust-proof.[17]

The question of churn design was part of the wider conflict between farmers and the railway companies. When some or all of the milk arrived in poor condition, the railway companies often blamed the large churn favoured by farmers, which they argued was unsuitable for its purpose, being heavy and awkward to load and unload from carriages; they advocated the adoption of a smaller and more manageable churn similar to the kind used in France. There was an economic dimension to this ongoing quarrel which belied the apparently straightforward question of what was the most practical design of churn. The railway companies charged farmers a rate per can for the return of empty churns, and if the farmers had agreed to use smaller churns, then the charge would have been increased for a given quantity of milk. This highlights the role of uniform practices of measurement and of standardized apparatuses of quantification in rendering possible the operation of formal equivalence across a territory. Thus, as long as the parochial system of measurement by churns was relied upon, changes to the established form of container were replete with implications beyond merely the possible improvements in functional efficiency. A system of measurement with the semblance of universality, sufficiently detached from local practice to enable it to function as the common currency of a network of practices transcending the local, had yet to be established.[18] This was perceived as such an impediment by agricultural reformers that a radical solution in the form of complete conversion to the metric system was being urged by some as early as 1906.[19]

The difficulty of establishing a uniform system was a symptom of the fact that the rationality of objective measurement did not already exist in some as yet undiscovered conceptual space just waiting to be put into operation, but had to be carved out and actively entrenched by modification of existing local practices and the relations they involved. That is why political efforts to enforce standard practices of measurement across the milk trade were not merely a technical matter but were also deeply political; they were a precondition for the transformation of milk into a marketable commodity on a national scale. It was not just a matter of displacing one system of measurement and imposing another. The problem was further complicated by the fact that existing customs of measurement varied widely between localities; there was no consistency. As one contributor to the *Journal of the British Dairy Farmers' Association* commented:

> In one district a given standard is in vogue; in another there is a different one. Here it is the barn gallon, there the imperial, and yonder the dozen quarts. Certain people are so accustomed to the first, or to the last, as the case may be, that they are helpless as children if they try to reckon under

> the one in the middle. And so it is that in districts where the barn gallon is the standard for milk, it would not be an easy thing to induce the people to voluntarily relinquish it. All the same, the barn gallon itself is illegal as a standard to sell by; only it seems to be nobody's business to call it in question.
>
> (Sheldon, 1893, pp. 28–9)

These practices of measurement were not only ingrained by habit, they were also closely bound up with the interests of the various groups involved in the wholesale milk trade; thus measurement was in no sense neutral and value-free.

Quite apart from the political role of standard measures in facilitating a national market for milk and establishing the national territory as a political and economic unit amenable to effective government, alternative methods of quantification offered distinct advantages to certain groups over others. In this way the very apparatus of objectivity and universality was shot through with partiality from the beginning.[20] The 'barn gallon' which was used in wholesale milk deals, for example, amounted to around 17 pints, but it was understood to mean a two-gallon standard 'with the odd pint thrown in as a make-weight'; extraordinary by today's standards, this 'odd pint' was regarded as a perquisite by the dealer.[21] Those who favoured retaining the barn gallon often contended that this 'odd pint' was paid for with the rest, but as one critic of the measure shrewdly observed, if this was so, then there was no logical reason why it should not be abolished and the barn gallon be taken as simply two imperial gallons. The fact was that farmers who sold their milk by the barn gallon effectively handed as a gift to the dealer a quantity of milk reckoned to be around 6 per cent of the total.[22] Unsurprisingly this created some fervent supporters of the barn gallon as a measure.

Whereas use of the barn gallon was technically illegal, the sale of milk by 'the dozen' or dozen quarts, was not, because unlike the barn gallon, the dozen was able to be translated exactly into imperial measures, being three imperial gallons. Nevertheless the dozen was very much local in character, and was not defined in law as a standard measure. The perceived advantage of the dozen was that it tended to enable prices to be cut finer than with the gallon, for the units it employed meant that it was resistant to the rounding up which often tended to accompany pricing by the imperial measure. Finer pricing, of course, was a matter of no small import for those involved, and there were many on both sides of the wholesale trade who strongly believed that they would be disadvantaged if the exclusive use of the gallon was strictly enforced. There was also the question of whether milk should continue to be sold by measure, that is by volume, on the wholesale market, or whether it should instead be sold by weight. While it was widely agreed that the weighing of milk was simpler 'in principle', the difficulty arose from the texture of the material, that is, from the particular material properties of milk. Whereas a gallon of water weighs 10lbs, allowing it to be the standard against which gallons of other liquids are calculated, a gallon of milk was found to weigh, on average, 10lbs 5ozs and a fraction. Not only was this a highly inconvenient measure, but it would have rendered the imperial gallon a useless standard for the sale of milk even before it had been fully established against the barn gallon and other local measures. Milk, unlike corn, therefore continued to be sold by measure, as the practical case for selling by weight was both confounded by the material properties of milk and subordinated to the political need to establish the imperial gallon as a standard unit.

Organizing time: networks of distribution

The railway companies assumed that the volume of milk being transported by rail would begin to diminish once the city herds had recovered from the immediate shock of the cattle plague and as the numbers of town cows began to return to their pre-1865 levels. There was therefore little concern with provision of the kind of arrangements which might have been considered necessary if it had been anticipated that in fact the town herds would never recover and that the milk traffic, far from diminishing, would continue to grow steadily in line with the population increase in the towns.[23] Hence no special trains were laid on for milk transportation; instead the churns were simply loaded into old second class carriages or 'vans' on ordinary passenger trains, sometimes with numerous other animal produce. Apart from the considerable practical difficulties which this involved in the regular loading and unloading of heavy churns from carriages not designed for the purpose, it also created sanitary problems, as one farmer complained:

> The van itself is in reality a travelling dairy, and it not infrequently happens that the said van is dirty and ill-ventilated, that along with the milk is carried hides and carcases of meat, and live fowls in baskets, while sometimes the atmosphere is redolent of fish.
>
> (Sadler, 1907, p. 19)

This was a constant source of frustration for farmers, who were just at this time beginning to feel the pressure of growing official concern for milk hygiene. Moreover, due to the ad hoc nature of the early railway milk trade and its lack of regulation, in cases where milk arrived at its destination in poor condition, it was overwhelmingly the farmer who was faced with the burden of having to prove that no malpractice had been involved, even in cases where the deterioration was clearly due to the manner in which it had been transported. Again this was a problem specific to milk and its particular properties; it presented itself as the question of warranty.

According to the principle of warranty, a retailer prosecuted in the course of his business for the sale of an adulterated article, if he could show that he had sold it in the same condition as it had been received, could escape being convicted of fraud. But there was a material difference between the application of this principle to different dairy products; in the case of cheese, for example, if it was found to be adulterated then there was no question but that the manufacturer was responsible, for once it has been produced the component parts of cheese cannot practically be altered. As James Sadler remarked, however:

> Milk, unlike cheese, is not a manufactured article (except in the sense that the cow is the manufacturer), and is not to any appreciable extent under the control of the producer, and no farmer, however clever he may be, can always ensure that the milk will be up to a given standard. The most he can do is to guarantee that when it leaves his hands it is genuine. After that of course, it can be adulterated by every person who handles it, and this can be done either designedly or through carelessness. That being so, no farmer should warrant his milk beyond the time when it passes out of his control. If he does, he is liable to be made a scapegoat for other people's sins, and the law as it stands enables this to be done.
>
> (ibid., p. 22)

This passage suggests that the legal difficulties over the application of the principle of warranty to milk arose from the sale of milk *as though it were* a manufactured good. In other words, milk was being treated for economic purposes as exclusively the product of human producers acting upon an inert nature, rather than as the hybrid co-product of both cows and humans and as much a 'natural' substance as a manufactured good.

As well as the legal issues, the use of passenger trains for milk transit impacted directly upon the temporal rhythm of the milk traffic, milk deliveries being restricted by the timetables of slow passenger trains, with significant implications. Indeed, both the timing and the frequency of milk deliveries were constrained by factors which were to a considerable extent nonhuman, namely the circadian bodily rhythms of the cow, and the action of microorganisms within the milk in contributing to its perishability. While the latter could be forestalled by the new refrigeration technology, at least enough to facilitate the transportation of milk by rail, it could not yet be obliterated, and the cow's bodily temporality had so far proved unsusceptible to effective manipulation. For decades farmers had attempted to find a way to reorganize the times of daily milking in a manner that would render the task less arduous and inconvenient, to no avail; every alteration of the twice-daily pattern had resulted in a notable reduction in the quantity of milk produced.[24]

There was therefore no alternative but that milk be despatched twice a day, as soon as possible after milking; every effort had then to be made to minimize the time between milking and consumption, by efficient transportation of the milk and particularly by ensuring minimum delays at each stage of its journey from the farm to the consumer. One distributor spelled out the problem at the Midlands Dairy Conference of 1909:

> Time is the very essence of the milk business, and as the morning deliveries to customers commence at 5 o'clock, or very soon after, there is not a moment to be wasted, and the same remarks apply to the later deliveries. The retail distributor realises that if he is to hold his own in these days of keen competition he must see to it that his rounds are served with almost clockwork regularity.
>
> (Horner, 1909, p. 40)

> There should be no despatching of the milk once a day, especially in the summer time, and of that still greater evil, the night's meal of one day with the morning's milk of the next day, sent to the station in the middle of that day to be despatched by the night's train, the one meal being 24 hours and the other 12 hours old before it leaves the departure station.
>
> (ibid., p. 38)

This was irreconcilable with the use of passenger trains, or indeed with the transportation of milk along with other commercial traffic, for neither of these was well adapted to the stringent requirements of the milk trade.[25] Thus there were persistent calls by farmers for the railway companies to provide special trains exclusively for the purposes of milk delivery. There were still no such trains in 1878; instead, milk vans were being attached to the fast long distance morning and afternoon trains.[26] By 1909, some of the railways were providing milk trains along the main lines, but there were still no milk trains on local lines.[27] Consequently, throughout the late nineteenth and much of the early twentieth century most milk still depended for at least a portion of its journey upon passenger or other commercial traffic and therefore risked considerable delays.

Different temporal modalities were involved in the movement of milk and the movement of passengers or other goods, the contradiction between them signifying the hybridity of milk and the tension between its human and nonhuman dimensions. On the one hand, milk was the produce of the cow, a nonhuman producer, and the relative inflexibility of milking times persisted as an umbilical connection to this nonhuman corporeality. On the other hand, milk was increasingly a modern mass-marketed commodity, an exchange-value, a manufactured good and thus a social and economic product integrated within the rhythms of urban life, industrial clock-time and the spatio-temporal flow of commodities. This tension expressed itself in the ongoing conflict between dairy farmers and the railway companies over all matters bearing upon milk transportation, which was almost uninterrupted from the late 1880s onwards.[28]

As the overseers of production, farmers were well acquainted with the inflexible nature of milking times and were aggrieved by what they saw as the intransigence of the rail companies in their reluctance to meet requirements which, after all, were 'objective', in that they flowed from the organic properties of the milk. In this view, though farmers were the visible 'actors', yet behind their agency and as it were 'pulling the strings' stood the 'nature' of milk itself. The railway companies, in contrast, whose representatives did not have the same direct experience of dealing with milk as a substance, were inclined to regard the demands of dairy farmers as excessive. Though they were prepared to make some concessions on grounds of perishability, they believed that milk, by and large, should be transported as would any other commercial product and they were reluctant to make the investments required to provide a service tailored specifically to the milk trade. This conflict was a critical factor in the emergence of producer organizations in the early twentieth century, as farmers belatedly accepted the need to act collectively in order to secure more suitable conditions of carriage for their product.[29]

As the railway network was extended and as the trade developed, the average distances which milk was carried increased substantially and the cost of bringing the milk into the towns grew steadily. The railway companies wanted to increase freight rates, ostensibly to cover the costs of the expanding trade, but farmers were determined to avoid paying higher rates for facilities which they felt were still inadequate or unreliable. Under the Canal Traffic Act of 1888 the railways began to try to raise the rates for the steadily growing traffic, triggering a dispute which was heard by the Board of Trade in 1890.[30] By this time several groups of farmers had begun to form local associations, in a move intended to resist any raising of the rates by means of co-operation and collective organization in pursuing their sectional interests. Thus was set in motion a dynamic in which conflicts of interest between the groups involved in the milk trade, and especially between producers and distributors, increasingly shaped the form of regional and national organization of those groups. This led gradually but inexorably to the centralization of control and the concentration of ownership which characterized the trajectory of the milk industry throughout the first half of the twentieth century.

Economic concentration and monopolization

The ad hoc distribution of milk from the railway platforms had by the close of the nineteenth century become dominated by a small number of large wholesale-retail companies, as the railhead distributors of the 1870s merged into the big milk corporations of the early twentieth century. The First World War led to further concentration of ownership, as competition between rival companies became increasingly difficult due to shortages, and as co-operation was introduced to meet national demand. In 1917, just one company controlled two thirds of London's wholesale and one-third of its retail trade after the four largest dairy companies in southern England merged to form United Dairies.[31] This dramatic concentration of control, though partly the product of systematic conflict between producer and distributor interests, quickly created the conditions in which increased co-operation between these groups became instrumentally rational; in a dialectical movement, the organizational response to this political-economic situation effectively transformed the situation and so demanded a new response.

There was a significant technological dimension to this, for as clean milk campaigns were stepped up and regimes of sanitary inspection intensified, dairy equipment previously regarded as the preserve of wealthy pioneers became an absolute necessity for farmers and wholesalers if they were to keep pace with the increasingly stringent demands of urban health authorities for milk which met the requisite standards of cleanliness. Producers and distributors who could not afford such expensive equipment as filtration plant, centrifugal machines, pasteurization facilities and proper cooling apparatus, were not only more vulnerable to inspection, but were also competitively disadvantaged, as rigorous standards of cleanliness became more and more central to the national strategy for mass marketing milk to wary urban consumers. Bottling facilities were also increasingly an advantage as the concern for sanitary arrangements was extended to the method of final delivery.[32] Hence producers, who had originally begun to form societies in order to protect their interests against those of distributors, wholesalers and railway companies, now used these organizations to begin to establish milk collecting depots near to railway stations, where milk was collected, refrigerated, and often tested for compositional standards and for TB.[33] From there the milk would be sent into town, where it was forwarded to a milk receiving depot, also typically equipped with refrigeration plant and testing and bottling facilities, before finally being sent out to retailers. There was some movement towards an extension of co-operation in both directions, as producers sought to master the problem of warranty by securing more control over transport and distribution, while wholesalers sought to minimize their risk of losses by gaining more effective regulation of supply.[34] Thus, for producers and distributors alike, collective organization and the pooling of resources in order to co-ordinate production and consumption and undertake serious investment in sanitary equipment became a key strategy for transforming the emerging sanitary regime into an opportunity for competitive advantage.

This concern with effective co-ordination of supply and demand and with the organization of the milk trade in such a way as to minimize risk and maximize security was not restricted to the dairy companies themselves, but was increasingly the problematic that framed government policy for dairy agriculture. The dominant economic paradigm underwent a remarkable transformation in this period as the government's long-standing reluctance to intervene directly in the economics of the milk trade was progressively routed by the argument that collective organization and planning on a national scale were indispensable to secure the most efficient, stable and secure organization of milk production, distribution and sale. It was in this changing political and economic climate that the use of what was called 'propaganda' to increase national milk consumption became conceivable. This shift was precipitated by the concatenation of various historical circumstances, many of which transcend the field of dairy agriculture, and most of which have been extensively considered elsewhere by agricultural and economic historians. I will therefore restrict myself to briefly outlining the political-economic context in which milk propaganda emerged.

The most crucial factor was the impact of the First World War, which meant that the national economy suddenly became viewed not merely in terms of wealth creation but also, very substantially, in terms of the security of the nation-state. It was clear that the outcome of wars between modern states would be determined by the strength of their respective economies, more specifically by their ability to adapt to the requirements of wartime munitions production and by their degree of self-sufficiency in essential foodstuffs. One typical report of the time described the situation as follows:

> The matter of producing the greatest possible quantity of food in this country has at last obtained the interest of the Government, and this interest has been forced upon it by the curtailment of imports due to the German submarine menace. The daily average loss of some 20,000 tons of shipping sunk by submarines, and utilization of a large proportion of our shipping for transport purposes, is reducing the merchant service so much that the people of the country are likely to have to depend very largely – at any rate during the war – on home-grown supplies.
>
> (Walker-Tisdale, 1917, p. 29)

It is unsurprising then that the Board of Agriculture became preoccupied with extensive comparisons between British and German milk production, as the primary axis of economic competition tilted away from financial rivalries between private companies and towards military-industrial rivalries between nation-states. Indeed, many articles during the war focussed upon reporting the results of various statistical analyses seeking to determine the relative productivity of the British and German herds. Their conclusions did much to galvanize those who wanted to introduce greater state coordination, political planning and collective organization of the dairy industry.[35]

It had previously been widely assumed that British milk production was more efficient than its German counterpart, both because British soils were known to be more favourable and because for the previous forty years the overall trend in British agriculture had been away from arable crop farming and towards stock farming, that is towards greater meat and milk production; whereas the trend in Germany had been something like the reverse.[36] However, even the most rudimentary analysis now revealed that German milk production was considerably in advance of that of Britain, the yield of milk per acre of cultivated land being 37.75 and 61 gallons respectively.[37] In part, this particular statistic could be explained in terms of historical differences in land use between the two countries, but this would not account for the persistence of the productivity gap under various other measures. The conclusion was inescapable; though Britain had seen significant improvements in dairy farming in the period since the late 1900s, it had largely relied upon the agency of private enterprise and the initiative of various agricultural societies, with minimal government assistance, for its modernization.[38] In Germany, on the other hand, there had been a deliberate and concerted effort by the state to increase national agricultural production by whatever means:

> With a view to becoming a self-supporting nation, prior to the inevitable war which its policy foreshadowed, Germany has been devoting itself very strenuously, to the development of its agriculture, with a result that during the past twenty-five years it has enormously increased its production of nearly all classes of agricultural produce.
>
> (Walker-Tisdale, 1917, p. 29)

> Technical and agricultural education has been developed in Germany now for a great many years, and that State had aided industry by spending large sums on development and employing tariffs to steady prices. Co-operation amongst farmers in Germany is very largely employed.
>
> (ibid., p. 32)

The result of this unpalatable discovery was that political opinion in Britain quickly became more favourable to a significant increase in state intervention and control, over agriculture in general, but over milk production above all. For milk was identified as the essential food *par excellence* on the grounds of its imputed status as an almost complete source of nutrition. At the same time, a secure and reliable supply of milk for the besieged populace became a central plank of national 'food security' policy, according to which it was deemed necessary to increase milk output at all costs. To this end, the National Milk Recording Scheme of 1914 was put into effect, and in 1917 the Ministry of Agriculture began to undertake a series of six-month-long experiments in 'cost accounting' at a large number of selected dairy farms. Officials would visit the farms on a weekly basis to check the accounts, make records, and amass more information with which to inform policy decisions on how best to increase milk production.[39] Such concerns for the security of the milk supply gave considerable momentum to an accelerating process of rationalization which was driven at least as much by political will as by economic processes.

These war-linked efforts to increase milk output independent of levels of demand tended to exacerbate a growing crisis of overproduction, as intensification and increased yields per cow failed to be matched by comparable increases in consumption per head. Encouraged by the state, the expansion of supply had become an end in itself, sealed off from the question of what effect massive surpluses would have upon the wider milk trade. During the war this disparity was largely invisible, due to endemic shortages and the structural distortions of a wartime economy, but the underlying crisis grew in severity, culminating in a serious collapse of prices after the war as the trade attempted a faltering return to its pre-war economic norms in what were by then very different conditions. Major fluctuations in prices were more or less persistent throughout the 1920s, which both antagonized consumers and led to growing labour problems, as farm workers became increasingly reluctant to take the brunt of the hardship and insecurity.[40] The crisis generated mounting pressure for some kind of government intervention to steady prices, increase demand and stabilize the industry. The Astor Committee, formed in 1917 to advise on the organization of the milk supply during the war, as well as suggesting measures to improve production, transport, and distribution, also discussed the pros and cons of government control, albeit without committing itself to the idea. The first attempts at collective price agreements between producers and distributors were made by the Permanent Joint Milk Committee in 1922. The following year, the Linlithgow Committee recommended further collective bargaining between producers and distributors as a method of dealing with the collapse of prices.[41] This pressure eventually culminated in the Agricultural Marketing Act of 1931, which paved the way for the formation of the national Milk Marketing Board (MMB) in 1933, a defining moment for organized marketing.

The Board consisted of fifteen elected representatives of producers and three government appointees; these were supported by twelve elected Regional Committees. Its sphere of operation was extended systematically and eventually encompassed not just the marketing of producers' milk, but also the keeping of National Milk Records.[42] The legislation which created the Board established at the same time a legal monopoly of the milk supply; all producers were thereafter obliged to sell their milk to the Board unless granted a licence to sell directly to a wholesaler. This allowed the government to control prices by becoming the universal intermediary between producers and buyers and setting up a pricing structure in which a 'pool' price was fixed each month.[43] Though at first independent sales made up a considerable portion of the national wholesale market, this portion was reduced throughout the following decade and the number of licences granted soon began to fall sharply as the Board tightened up its monopoly.

To return to the question of commodification, then, liquid milk was rendered commodifiable in spite of the resistances of its materiality through a multitude of technical and political devices. Increasingly insulated from natural or reversible time by the technologies of refrigeration and railway transportation, it was similarly insulated from the fluctuations of the market in order to secure the stable conditions required for planned increases in production and consumption. In this way the conditions of its economic viability were created by means of massive and sustained intervention; milk had indeed become a modern mass marketed commodity, but this commodification was the product of myriad micro-political techniques and institutional fixes rather than abstract or autonomous economic forces.

'The backbone of Britain': milk security and milk propaganda

It was in the context of moves towards national control of the milk supply that the use of propaganda as a tool to stimulate consumption and manage levels of demand began to assume critical importance in the early 1920s. The National Milk Publicity Council (NMPC) of England and Wales was formed in February 1920. Its reason for existence was to organize a national publicity drive to raise overall levels of milk consumption.[44] Sanitary discipline and surveillance had by that time so permeated milk production and handling that the product was becoming technically safer for human consumption than it had ever been, but the perception of milk as a risky and potentially contaminated food, and especially its association with tuberculosis, had not yet been excised from the public imagination; this was identified as a key reason why milk consumption per capita had not increased in line with the growing safety of the milk. Other reasons included oscillating prices, which national control was designed to stabilize, and unreliability or insecurity of supply, a matter that the war had brought to a head and which improved transport and distribution networks, together with increasing coordination of supply and demand and an extension of collective organization, were believed to be within reach of resolving. The modern railway distribution network, along with collection and distribution depots increasingly making use of sophisticated pasteurization, refrigeration, testing and bottling facilities, meant that the physical and organizational infrastructure to sustain mass consumption now existed. It was therefore believed by reformers in the industry that if prices could be stabilized and the public could be convinced that milk was now safe beyond question in terms of hygiene, cleanliness, then the final barriers to mass consumption would be eliminated and levels of consumption per head would soon increase accordingly.

This was based of course upon the assumption that people already *wanted* to drink more milk but were discouraged by factors of hygiene, supply, or price. Indeed, all of these supply-side measures only succeeded in creating a situation in which people *could* drink more milk, and did nothing positively to persuade people that they *should*. Thus, while the technical and economic aspects of the market were being fine-tuned and micro-engineered to facilitate increased consumption, the motive force behind consumption, the actual desire for the product, the need or want itself, was still generally assumed to be beyond manipulation. Though a burgeoning market in print advertising since the 1850s had begun to recommend an increasing array of goods to the middle classes, it was the consumer rather than the advertiser who was still the primary agent in this early strategy;[45] the increasing affluence and commensurate material acquisitiveness of the Victorian consumer were served rather than generated by the explosion of advertising. The prevailing common sense, quite simply, was that people could not be *made* to want things, and that the function of advertising was to inform consumers of the availability of certain goods, or to promote particular brands, rather than to engineer a previously non-existent desire. This paradigm would not outlast the 1920s.

The formation of the NMPC marked a significant point of transition, for unlike the pioneering private milk advertising that had preceded it, which largely consisted of the promotion of particular brands in products such as condensed or powdered milk to existing consumers, the Council now co-ordinated a national campaign promoting the consumption of milk in general and targeting the entire potential market. The objective now was not merely to sell a specific product but to shape behaviour on a grand scale, its stated aim being to *make* people drink more milk, a task which for the first time was defined explicitly in positive terms as the active creation of new wants. At the centre of this was a newly systematic emphasis upon the nutritional value of milk.

The notion that milk is a uniquely nutritious food was in itself nothing new, having been espoused in one form or another by those whose interests were tied up with the promotion of milk for over a century, but at the same time it had always been vigorously contradicted by popular discourses, particularly in the towns. These claimed both that milk was almost invariably contaminated and that even if its purity could somehow be assured it was suitable only for young children and not for adults.[46] This was largely because milk was still strongly associated with breastfeeding, which is to say with the role of human milk, with nurturance and with the bodily relationship between women and babies. Hence there was an ingrained perception that drinking milk was somehow not masculine, that it was 'soppy' even, as the word 'milksop' implied. Other critical discourses on milk had emerged from the experience of widespread adulteration, from the watering down which in the 1850s and 1860s was perceived to be almost universal, from the infant mortality associated with cow's milk, and from the very poor general standards of hygiene that had typified the urban milk supply in the years preceding the Cattle Plague. There were even tales of fish swimming around in milk which was waiting be sold to the consumer, an image that epitomizes the way in which milk had become a key symbol of the public health threat posed by unreliable and contaminated food. Reinforced by the recurrent milk 'scandals' that appeared in the papers every time there was an accusation of malpractice, this image captured the popular imagination and became the dominant representation of milk in the towns for much of the mid- to late nineteenth century.

As successive legislative measures together with changes in the structure of the milk supply after 1865 reduced the incidence of adulteration, the popular perception of contaminated milk was gradually replaced by a new anxiety over tuberculous milk, as impurity gave way to a different kind of contamination, namely bacterial infection. This was both compatible with technically 'pure' milk, and propagated not by devious human agency but by an apparent conspiracy of different kinds of *non*humans. The legacy of these popular anxieties was that the twentieth century inherited the perception of milk as an inherently risk-laden food. Nor was this limited to popular discourse. The Medical Officer of Health for Salford, for example, was a major critic of milk and what he saw as its detrimental public health consequences. In 1925, he launched a series of 'health weeks', which were attempts to 'educate' the public on the risks of contaminated, adulterated and 'impure' food.[47] Milk featured centrally in this campaign of what he called 'Public Health Propaganda', echoing the interventionist language of the National Milk Publicity Council:

> There is a clear case for Public Health Propaganda – the spreading of the Gospel of Hygiene and the inculcation of a 'Health Conscience' . . . Health Weeks and lectures will appeal to a certain class of people. The cinema has proved a very valuable educational means, as has also wireless broadcasting. The distribution of health leaflets and the publication of articles on health matters in the newspapers are largely employed and constitute useful propaganda.
>
> (Osborne, 1926, pp. 79–90)

As part of this campaign there were permanent displays in the ground floor windows of the Health Department on Salford's Regency Road from 1925 to 29. These were the subject of considerable public interest at a time when urban space was nothing like as saturated with commercial images as it is today and when people consequently were far less jaded and much more impressionable in their attitudes to visual publicity. One of these displays was entitled 'Pure Milk – The Perfect Food' but in fact stressed the dangers of 'impurity' and showed the organs of animals injected with tuberculous milk in dissection cases.[48] In 1927, the MOH staged an exhibition of negative milk publicity, which he called an 'Exhibition of Impropaganda', displaying various claims concerning the impurity of milk. This campaign culminated in December 1928 in a two-week-long exhibit organized by Salford Health Department, which reportedly attracted 30,000 visitors, and which included displays on 'pure milk' and 'food and food adulteration'.[49] Similarly, three years earlier, there was a dispute between the NMPC and the Chairman of the Food Education Society, Charles Hecht, who added his scientific opinion to the popular notion that milk was 'a food for babes not men'.[50] The NMPC, however, together with the National Institute for Research in Dairying and, later, the MMB itself, in conjunction with the Milk Nutrition Council and other bodies, set about systematically drowning out these dissenting voices by means of an increasingly organized, well-funded and far-reaching propaganda campaign, which sought to establish the discourse of 'uniquely nutritious' milk as the universally accepted scientific truth.

At the centre of this was a strategic effort to link the consumption of milk inextricably with the welfare of children, by making it synonymous with healthy physical and mental development and well-being. To this end, the NMPC sponsored a 'milk experiment' at Allcock Street School in Digbeth, a particularly poor area of Birmingham, conducted in 1923 by Dr George

Auden, the School Medical Officer. This involved giving thirty boys and girls aged from 7 to 11 an allowance of one pint of milk daily for two months and noting weight and haemoglobin increases against a control group. It was duly concluded that 'not only did milk foster physical health and development, but it also had a direct effect on scholastic ability'.[51] This was one of many similar projects in which the inter-war emergence of the modern social welfare paradigm was linked to a growing concern with the collective bodily health of the citizenry, which came to be viewed in terms of the health and security of the body politic. Thus the health of the population had increasingly to be known, measured, recorded and mapped, in order to facilitate its effective government. At once political and scientific, this incipient 'bio-politics' was also bound up with commercial interests, for if people were to be told that they *should* drink more milk and if its increasing availability were to be steered into the desired patterns of collective behaviour, then a didactic element was required as a disciplinary mechanism. This was to be found in the emphasis on health, which modern medical discourse had already begun to infuse with moral undertones, particularly the health of growing children.

Typically, the suggestion was that 'good' parents, who cared about the development of their children and who wanted to give them the best possible start in life, ensured that they drank plenty of milk, which, it was repeatedly claimed, was incomparable to all other foods in its nutritional value. It was the task of nutritional science in the service of the milk industry, the interests of which were increasingly conflated with those of the public at large, to embed this idea as deeply as possible within the popular imagination.[52] In addition to the Birmingham study, doctors H. C. Corry Mann and J. Boyd Orr carried out large-scale 'experiments' throughout the 1920s which purported to establish the nutritional value of milk. These involved measuring the effect of regular milk rations upon the height, weight, and 'intelligence' of groups of schoolchildren from different regions over several years. The findings were published in a Medical Research Council Report of 1926 entitled *Diets for Boys of School Age*, in time for a Pasteurization Conference in London which was addressed by the Minister for Health, Neville Chamberlain.[53] Of course, this latest study confirmed the results of the Digbeth experiment, and was instrumental in the subsequent adoption of free school milk as Labour Party policy.

In 1928, Mann published another report, similar to the earlier longitudinal 'experiment' except that the evidence in this case, he claimed, was based upon a case history of twin brothers who were separated at birth, their common paternity standing in as a kind of crude scientific control. One of the brothers was sent to a public school where he reportedly obtained 44–51 per cent of his animal protein from milk, the other to a poorer school where the staple diet was bread and margarine with very little milk and where around 13–15 per cent of his animal protein was obtained from dairy foods. According to the author, the first boy, who had been 'exceedingly sturdy' in boyhood was, at 25, 'abroad in Government service, a healthy and well developed man', while the other boy 'became a clerk in the city of London and at twenty-five was still under-developed'.[54] It need hardly be said that the 'scientific' credentials of both this and the earlier 'milk experiments' were doubtful, their methodologies being highly questionable.

In this last case, for example, the report proceeds on the prior assumption that the *percentage* of the total protein consumed which is obtained from animal sources, and from dairy foods in particular, is the key variable; it makes no direct comparison of average daily calories or protein intake *per se*. Obviously enough, if one boy generally consumed an adequate quantity of protein and calories while the other boy's diet was notably deficient in these, then the results described are only to be expected; one boy was malnourished, the other was not. It does not follow that this has anything special to do with milk, except insofar as it happened to be the source of much of the extra calories and protein for the well-nourished boy, which could well have been obtained from other foods. Indeed, throughout these 'experiments' of the 1920s and 1930s the nutritional value of milk, far from being established 'scientifically', was typically affirmed rhetorically by means of a skilful arrangement of 'the facts'. This is unsurprising given that the main purpose of these experiments was to provide a source of positive milk publicity rather than to meet stringent scientific standards; they were 'propaganda for consumption' which had merely adopted the discursive form of scientific investigations.[55]

What is most interesting is the class dimension of these 'milk experiments', how they sought to exploit the aspirations of parents for their children by symbolically linking the consumption of milk with notions of class mobility. In the case of the separated twins, for example, the first brother is sent to a public school and enjoys all the associated privileges and advantages, one of which is an adequate diet; whereas the second brother is sent to a relatively impoverished school and enjoys none of these benefits. Moreover, the first boy ends up 'serving his country abroad', with the connotation that as 'a healthy and well developed man' he is 'fit' to represent his nation 'on the world stage'; whereas the other, milk-starved boy becomes a clerk, a career of relatively lower status. The full rhetorical impact of the contrast hinges upon contemporary notions of empire linked to national service and is emphasized by the contrast between the overblown language used to describe the first boy's career and the flat statement of facts used with regard to the second boy's occupation. Moreover, while one boy is 'active' in the service of his country, the other boy by clear implication is 'inactive', for he 'sits' in an office all day. By this sleight of hand the boys' differential physical development is subtly linked to their different careers; we are invited to believe that the second boy has failed to reach his full potential because he perhaps does not have the bodily and mental energy of his brother. Thus, while the main determinant of their diverging careers may be identified sociologically as the difference in their class backgrounds and schooling, there is nevertheless a carefully connoted relationship in this narrative between social achievement and the consumption of milk. There is no need for a causal connection to be established, a lingering association was enough to shape behaviour. Whatever its factual accuracy therefore, and given the nature of the case, there are certainly grounds for scepticism. This kind of scenario served as an allegory for the impact of social class upon both bodily health and social position, into the centre of which milk was carefully insinuated as a symbol of both physical and social development, which in turn were codes for social class mobility.

'Social' milk: political technologies of milk marketing

Once the health, physical and intellectual development, and social achievement of children had begun to be symbolically linked to milk, the industry had established a strong foothold from which to expand its propaganda campaign. Children meant families, and if milk could be marketed as a product for the whole family, rather than just something for young children, then it could be made necessary to all sections of the population. The strategy of marketing to families was therefore seen as the most effective route to mass consumption. In his retrospective tribute to the work of the National Milk Publicity Council, Alan Jenkins put it as follows:

> Workers, women, mothers, children, sportsmen, caterers, nurses, doctors – all were targets for propaganda. Posters showed boys carrying off trophies at school sports – 'Mother I'd win if I drank milk!' 'Milk in the Home – What Doctors Say'. A girl in a cloth hat was shown in a comfortable middle class café, with a clean table cloth (and probably a palm-court orchestra out of sight) saying to a nice, starchy waitress – 'and a glass of milk please'. 'To keep your looks and your health, Drink MILK.' There were advertisements, too, on the Underground.
>
> (1970, p. 88)

In the year preceding August 1925, the NMPC gave 2,614 lectures to an estimated audience of 250,000 people, and distributed 852,000 leaflets. At the same time £5000 was spent on colour posters, using slogans such as 'Milk – the Backbone of Britain' and 'Fresh Milk Makes the Ideal Baby'.[56] Advertising properly so-called, however, was still considered highly expensive for the purpose, and it was not until the formation of the Milk Marketing Board in 1933 that sufficient funding was found to ensure that milk advertising became extensive. Meanwhile the NMPC put 'public relations' at the core of its propaganda campaign, much of which was local in character, and which by connecting milk with the fulfilment of various social roles, particularly within the family and in the service industries, succeeded in submerging the industry's commercial imperatives under an apparently higher national purpose. Milk was identified not just with the health and vitality of children, mothers, fathers, manual workers, nurses, doctors, and so on, but therein with the health and vitality of the nation at large. Accordingly the NMPC forged connections with Scouts, Guides, Welfare Centres, Adult Schools, Women's Institutes, the Salvation Army, Conservative Associations, Mothers' Unions, and Temperance Societies. The impact of all this 'public relations' can be gauged by the statistics of the four largest rail companies; in 1923, 72 million gallons of milk was brought into London by rail, but by 1927 the figure was 117 million gallons, an increase of over 60 per cent.[57]

In 1927, the NMPC instigated a school milk scheme under which a third of a pint of milk could be purchased by children for a penny; this was subsidized by the government, and children from poorer families received their milk gratis. By 1931, 800,000 children were making use of these school 'milk clubs'. By 1939, more than 50 per cent of children in state-aided schools were receiving subsidized milk.[58] It was not only schools that were used to encourage people to drink more milk, however; under the 'Milk in Industry' scheme, workers in primary industries and manufacturing, as well as office workers, were encouraged to drink milk during their breaks. An experiment to measure the effect of milk consumption upon absenteeism was even suggested. Although at the time this was deemed too elaborate and expensive an undertaking, a scheme along the same lines was eventually put into effect some decades later.[59]

These kinds of schemes were effective both ideologically and practically; on the one hand, they reinforced a strategic association between milk and social welfare, while at the same time using subsidies to help to inculcate people into the habit of drinking milk, especially while young. They were therefore calculated investments designed to produce not just the consumers of the moment but the markets of the future. Meanwhile didactic 'milk experiments' continued to produce a steady stream of apparently authoritative evidence in support of the nutritional importance of milk. The National Institute for Research in Dairying published four reports on 'the effects of dietary supplement of pasteurized and raw milk on the growth and health of school children', the last in 1939. The first report dealt with an experiment sponsored by the Milk Nutrition Committee, comprising representatives of government departments, the Medical Research Council, the Milk Marketing Board, and other bodies, which claimed that children in the 8–14 age group who were receiving supplements of milk not only showed greater gains in weight, height, and chest measurement, but also, apparently because of the general improvement in their health, obtained more benefit from their lessons.[60] This theme was driven home relentlessly and corroborated by numerous similar experiments in England and Scotland throughout the 1920s and 1930s.[61] Thus, in an age highly deferential to scientific authority, the organizations centrally responsible for milk marketing were increasingly able to cite scientific 'evidence' in support of their claim that milk was a uniquely nutritious food. By the outbreak of the Second World War this had become indisputable through sheer force of repetition; it was a matter of common sense.

The creation of a mass market for milk was not just a matter of modern distribution networks and the deployment of propaganda, however, it was also a question of knowledge; the potential market had to be known in detail in order to be effectively manipulated. It was recognized that the market was not a homogeneous entity, but the objective was to make it as homogeneous as possible by targeting the different groups which it comprised and eradicating the differences between them in this one aspect of their everyday practice. The aim was therefore to make the heterogeneous market *behave* like a mass, to make *everyone* drink more milk; this demanded propaganda tailored to specific social groups. People had to be known in their difference precisely in order to be de-differentiated in their consumption; mass production of a standardized product demanded standardized consumption. The corollary was that not only the product but the consumers themselves were to be mass 'produced' and the raw material for their production was official statistical knowledge.

Attempts were made to chart and measure differential milk consumption according to social variables, with a particular emphasis upon class and income differences. The intention was to quantify as far as possible how much scope there was for increasing the consumption of milk among the different social classes:

> Orr (1936) finds that the consumption of milk varies from 1.8 pints per head per week in the lowest income group (10*s*. per head per week) to 5 pints in the highest income group (45*s*. per head per week). In order to adapt the level of milk consumption of the lower income group to that of the higher income group, the milk consumption of the whole country would have to be increased by 80%.
>
> (The League of Nations, 1937, p. 115)

This strategy was not unique to Britain, as is clear in a League of Nations bulletin of 1937:

> Another important factor influencing milk consumption is the number of members in the family, In general, it has been found that the larger the size of the family, the less the *per caput* consumption. This is evident form the results of investigations in various countries. For example, we may quote the enquiry carries out by Burns (1933) in the County of Durham (England) into the milk consumption of a thousand poor families making use of child welfare centres.
>
> (ibid., p. 114)

> The right plan in any country is to draw up a general nutrition policy, based on a thorough knowledge of the distribution of income and on the amount spent on food in the different income groups. The place occupied by milk in any such policy can then be carefully defined.
>
> (ibid., p. 116)

In this way the attempt to draw up a socio-economic map of differential milk consumption was part of a grander political objective, which was to regulate the nutritional profile of the population according to a general plan, to move towards a 'rational diet' for the nation. In the means by which this was pursued there was more than just a conceptual link between the 'economy' in the national sense and 'economy' in the sense of the budgetary planning and organization of a household; effective management of the former was held to depend upon objective knowledge of the latter, which had therefore to be studied accordingly: 'Any nutrition policy designed to secure a *rational diet* for the *population as a whole*, and to adapt the consumption of milk and milk products for the requirements of health, must be based in a detailed study of family budgets' (ibid., p. 130). It was found both that families in the lower income strata consumed less milk, and that milk consumption per head declined markedly as the number of children in the family increased.[62] The conclusion was drawn that in order to encourage greater milk consumption by relatively underprivileged social groups, milk welfare schemes had to be put in place to provide subsidized milk rations targeting those 'most in need':

> So long as the relation between prices and the purchasing power of various population groups renders it impossible for everyone to obtain sufficient milk and milk products, so long will social measures be necessary to ensure a greater consumption of these articles, particularly by pregnant and nursing women, infants, children of pre-school age and unemployed or indigent persons. Among the various measures to increase milk consumption, we recommend the grant by the State of subsidies for the provision of milk to schools, to pregnant women, nursing mothers etc.; facilities for the provision of milk in factories, offices, restaurants, railway stations etc.
>
> (ibid., p. 131)

In its linking of statistical knowledge of the population with measures to increase consumption across the board, this strategy paralleled the rationalization that was taking place simultaneously in the sphere of production, in cattle feeding and breeding practices, where official milk records were increasingly being adopted in order to increase productive yields across the cattle population. In both cases, statistics were used as a disciplinary mechanism, to enforce a uniform rationality upon a field of practice by a process of aggregation of the individual bodies involved, rendering them numerical abstractions, quantitative units. This enabled those units falling below a certain level of either production or consumption to be identified, and thence either physically eliminated or behaviourally modified. Such political technologies crossed, permeated and ultimately micro-constituted the fields of production and consumption as such, which were being subjected to similar processes of rationalization. This was by no means an abstract, autonomous or self-moving process, but was the cumulative effect of an ensemble of technico-political machinations driven by a historical conjunction of interests between producers, distributors, local authorities and the state.

After its foundation in October 1933, the Milk Marketing Board became responsible for advertising and direct promotion while the NMPC co-ordinated other forms of publicity. The Board now spearheaded efforts to develop the market; it sponsored the new 'milk bars' which began to be established, further developed the Milk in Schools Scheme and expanded the Welfare Milk Schemes in depressed areas.[63] Throughout the 1930s, the Board was spending an average of £60,000 a year on advertising. Propaganda began to be organized and concentrated on a regional basis, with the NMPC holding 'milk weeks' in selected towns, during which all forms of media were harnessed to ensure the maximum impact by means of posters, displays, leaflets and local press publicity. In addition to government grants and subsidies, by the 1930s, the NMPC was able to draw upon a budget of around £80,000 a year collected from producers and distributors.[64] Among those contributing both funds and members to the Council were not only the British Dairy Farmers' Association and the National Federation of Dairymen's Associations, but also the National Institute for Research in Dairying, the Sanitary Inspectors Association and the Society of Medical Officers of Health.

This underlines the organizational convergence of the milk industry with the public bodies which were responsible for regulating it. Indeed, the conflict between sanitary authorities and commercial milk sellers that had characterized the early modernization of the industry was gradually being replaced by something more closely resembling an alliance. In part, this was because a certain basic standard of sanitation had become widely practised, but it was also due to the accumulation of capital and the growing power of the large milk corporations, together with the changing structure of distribution, and the government's interest in stimulating milk consumption in order to ease the growing problem of overproduction. All this meant that the industry was increasingly in a position to enlist the support of the 'experts' with which it had once battled, a situation which was symptomatic of the fact that increasing the consumption of milk had become a central symbolic and strategic objective of the emerging national welfare paradigm. Thus the remit of public health officials was no longer to protect people from adulterated, contaminated or infected milk, but to ensure that they drank it in sufficient quantities.[65] It was no longer milk but its absence that was regarded as a social problem; the propaganda war had been won.

Conclusion: commodities, the 'economic', and the nonhuman

The late nineteenth and early twentieth centuries saw a decisive transformation of the milk trade, as local networks of production and consumption were developed into or replaced by an increasingly integrated national system of transportation and distribution, together with the management of demand through organized marketing and the national control of prices and supply.

The agency behind these changes included not just that of the state, successive governments, private dairy companies, agricultural associations, producer organizations, retailers and wholesale conglomerates, but also that of railway companies, sanitary inspectors, health authorities, nutritional scientists, advertising agencies and consumers. Indeed, when one traces the commodification of liquid milk in terms of its overlapping material, technical, political and discursive conditions of possibility, it is clear that 'production' neither determined 'consumption' nor vice versa, for these terms are abstractions referring to general processes and the networks of practice encompassed by these abstractions are too heterogeneous to be so generalized.

What *is* clear is that producer organizations were involved in an ongoing negotiation, characterized sometimes by conflict and sometimes by cooperation, with successive governments, wholesalers and distributors, to engineer a distribution network and a marketing system in which levels of consumption could be managed and made more durable in the interests of production. The networks were shaped in such a way as to facilitate the reproduction of their own conditions of existence. In this limited empirical sense, without drawing any overarching theoretical conclusions, it is reasonable to say that the autonomous action of consumers had very little to do with the emergence of modern levels of milk consumption. The means by which this consumption was produced are better understood in terms of the material and discursive tactics involved, the techniques-in-use by which physical problems were overcome and things and bodies governed, the discursive shifts through which new ontological boundaries and identities were inscribed upon phenomena, and the relationships of power, knowledge and interests implicated in these techniques and discourses.

Such an approach suggests that the existence of commodities is not contained within a separate sphere called 'the economic'. On the contrary, commodities imply an elaborate network – material, technical, epistemological, political – in short, substantially extra-economic in character, which must be in place in order for markets to operate; hence economic relations are emergent properties of technico-political networks.[66] Indeed, if one declines to take at face value the concepts of 'production', 'consumption', and the 'economic' as such, it is possible to reconceptualize these in terms of the historical networks of practices, techniques and relations which micro-constitute the field of commodity exchange.[67] Having gone some way towards deconstructing 'economic' categories in this way, it becomes conceivable to go further, to question the conceptually sealed domain of an exclusively human 'social production', of a 'socio-economic' realm, hence to move towards a 'symmetrical' or non-anthropocentric account. From this point of view, it emerges that there were other significant factors at work in the transformation of milk into a commodity, other 'agents' even, which had hitherto been marginalized only because our humanist ontological and epistemological assumptions had precluded the possibility of their acknowledgement as 'actors'; these include cows, milk itself, and micro-organisms. Thus, when the deconstruction of the 'economic' makes a symmetrical standpoint thinkable, one can begin to reckon nonhumans into the historical constitution of the 'economic', from whence it is a short step to do the same for the 'social'.

To do so, however, requires a historical methodology which eschews explanation in terms of trans-historical subject-agents and pre-constituted domains of objects, and which insists instead that the processes of constitution of such subjects, agents and domains are the very thing that must be accounted for. A social and economic history or historical sociology as such takes the primacy of the human subject–agent for granted. What is needed instead is a symmetrical, genealogical approach, which dispenses with the privilege usually accorded to the human agent and which acknowledges the role of nonhumans in history. By so problematizing preconceived notions of what constitutes an agent and what 'social reality' consists of, it is possible to reckon the relations between humans and nonhumans into the heterogeneous networks which underpin the historical record. One may then posit that milk as a commodity is not simply the product of human relations of production, of wage labour, and of socio-economic actors, but is constituted through a separation of human from nonhuman producers; it is brought into existence as a 'social' and 'economic' object through less visible processes wherein its specific nonhumanity is systematically effaced. I want to argue in fact that this work of purification was, and indeed continues to be, an ontological condition of possibility of the expression of milk as an exchange-value, as a commodity.

Once this is contemplated then a rather different and less visible history emerges, a genealogical history which traces the technologies and processes of enactment that have created the conditions for commodification. This history identifies the key moments in the development of the modern dairy industry not in the quasi-causal impacts of the railway expansions of the 1870s and the propaganda campaigns of the 1920s and 1930s, but in the ontological politics of the 1865–66 cattle plague and its regulation, and in the politics of knowledge of the struggles against milk-borne tuberculosis in the 1890s and after. It is to this other history of milk that I now turn.

3 Culture, order and disease in late nineteenth-century British dairying

> Disease is at one and the same time disorder – the existence of a perilous otherness within the human body, at the very heart of life – and a natural phenomenon with its own constants, resemblances, and types.
>
> (Foucault, 1970, p. xxiv)

Zoonotic contagion: an ontological disorder

Disease is usually thought of as a thoroughly natural phenomenon, hence properly the object of biological science and pathology. Insofar as it is allowed that disease has a social dimension, this is usually strictly separated from its biological, natural aspects. On the one hand, we have the sociology of health, on the other, epidemiology, each sealed into its own conceptual domain. Thus one may study at will the relationship between health and social class, analyse the changing discourses around disease and its prevention, or study the effects of particular policies, but as soon as one entertains the notion that disease *itself* may be a social phenomenon, rather than merely a natural one which is relationally distributed, discursively understood and politically managed in deeply social ways, then one quickly encounters a yawning chasm revealing the conceptual limit of social science, the self-imposed horizon of its modernist imagination.[1] But what could be more social than disease? To be sure, disease is every bit as collective as language, indeed more so, for it speaks the universal language of matter, recognizing no linguistic boundaries and excluding no-body from its community. But if disease in itself constitutes a social relationship, quite over and above the social relations *of* disease, then it is a relationship that profoundly challenges our inherited conception of the social. It is a material relation between bodies, but one which elegantly undermines the Cartesian notions of human meaning and agency that so often underpin sociological discourse *on* the body, leaving us with a corporeality that is still stubbornly social but which refuses to be reckoned back into some form of idealist, linguistic, semiotic, in short narrowly 'cultural' conception *of* the social.

This is distinct from the notion of disease as *metaphor*, which is one way that the sociality of disease has been conceptualized by anthropologists;[2] my point is not that disease has cultural and metaphorical significations, a claim which hardly upsets the customary dualism of social science, but that disease *itself*, in its very material dynamics, is thoroughly and immediately social. Thus understood, disease compels us to conceptualize sociality without tacitly circumventing recognition of its inherent materiality on its own terms. At a stroke, sociality seems to rise up before us and cast off the dualist categories which for so long we had relied upon to contain it. Disease then is deeply problematic, for it fundamentally transgresses the great divide between mind and body, subject and object, culture and nature, human and nonhuman, which is the ontological heart of modernity, its knowledges and disciplines. More than just a practical threat to human life therefore, disease poses a threat, at once material and ontological, to modernity, its idea of 'culture', and its very notion of 'humanity'.

No kind of disease more dramatically destabilizes the dualist categories of modern humanism than a disease which is communicable between animals and humans – a zoonose. In its transmission from nonhumans to humans the zoonose threatens the ontological security of the human domain by its blurring of the anthropocentric boundary between 'culture' and 'nature'. The very existence of the zoonose materially undermines the coherence of the notion of 'culture', which is predicated upon a discursive and actual separation of an exclusively human domain from the nonhuman world. It is this ontological separation that effectively constitutes 'culture', and constitutes it as the secular translation of a divine or sacred realm, in which human action and meaning hold sway and from which history is born. Indeed, just as the modern secular sphere emerged from thoroughly religious roots, so this notion of 'culture' involves an unmistakeably theological conception of humanity.[3] By divorcing 'culture' from 'nature' and claiming it exclusively for humans we have managed to tacitly retain the notion that human lives are essentially affairs of the spirit, that humans, in other words, have souls. On this basis alone it becomes possible to contrast human lives with the purely bodily, material lives of nonhuman animals, which must live outside of the sanctified sphere of 'culture'.[4] Just as nonhumans were denied entry to the community of souls in the neo-Aristotelian Catholic theology of St Thomas Aquinas, so they are denied entry to 'culture' – that modern community of souls – in the secular theology of humanism.[5] Animals nevertheless continue to occupy a marginal and therefore dangerous position in relation to humanism for they are deeply hybrid, refusing to be either of 'nature' or 'culture', and thus they present a unique challenge to the coherence and purity of these meticulously constructed categories.[6] It is not enough to say that animals fail to fit neatly into these binary domains and so constitute an anomalous set of cases, leaving the categories intact; rather, animals show us the fundamentally tenuous nature of the categories themselves.[7]

Nonhuman animals are therefore the irresolvable problematic which humanist discourse must somehow suppress and contain; hence the relation between humans and animals is something of an absent centre for the humanist worldview. Only by rendering animals unambiguously part of an imagined object-nature, hence purifying the divine human domain of subject-culture, not just in discourse but in material practice, can the binary ontology of humanism be sustained. In relation to this ontology, zoonotic disease is the great leveller, proclaiming the radical equivalence of animal being, showing that everything takes place on the same level, and rendering visible the Christian 'Great Chain of Being' lurking beneath secular humanism.[8] In its indifference to the dualism of human and animal the zoonose is a dangerous hybrid, an ontological disorder; its existence articulates a kind of agnosticism of form, casting unforgiving light on the secularism of modernity and exposing it for a specious secularism of content thinly wrapped around another theology. Like all violations of articles of faith, therefore, the zoonose must be carefully censored; it demands a work of purification which is at the same time a work of discipline and an imposition of order, at once social and material.

Humanism therefore labours to re-establish the binary divisions which sustain the anthropocentric purity of 'culture', in this case by containing the zoonotic disease itself, which threatens to blur such divisions materially. This is not of course the explicit aim of the various institutions, knowledges and techniques involved; their conscious purposes are more immediately practical, but the discourse frames such practices as disease control while at the same time these practices instantiate its categories. Thus humanism constitutes the overarching regime of truth, the taken-for-granted organizing paradigm, which renders the zoonose not merely an economic problem, nor simply a problem of public health – though undoubtedly it is both of these and therefore a socio-political problem – but simultaneously an ontological, indeed, a theological problem for modernity.[9] Political in the sense that its instruments are those of official regulation, of disciplinary enforcement, of scientific and administrative knowing, humanism operates through forms of knowledge, rationalities, and the management of bodies in space, by constituting objects and domains, defining what can be known, what can act upon it, and by what means. It works 'horizontally' across social power relations, constituting their epistemological ground and extending its reign inside their networks of domination.

In another sense, the politicality of humanism is unlike what is usually thought of as the political, since its objective is not the domination of one human social group by another, although this may be involved, but the definition of what lies within the domain of the human, the social and the cultural, and what does not; it is an ontological micropolitics which is the condition of possibility of politics proper. Its object is the entrenchment and perpetual reproduction of a specific ontological division of the world, which constitutes and legitimizes a specific mode of species relations. Thus the sanitary discourse and veterinary pathology incipient in the latter part of the nineteenth century, which attempted by an array of increasingly exacting measures to contain animal disease, both zoonotic and epizootic, were the disciplinary elements in an ensemble of technico-epistemological practices serving to police the material and discursive boundaries of the human; taken together, these were constitutive of a political technology of humanism. To be clear, this does not imply that the discourse of humanism is in some way hidden beneath the surface of social life. Nothing is hidden except in terms of the sort of ontologically layered and stratified picture of 'society' to which humanism and its disciplines give rise. On the contrary, humanism, as I understand it, is thoroughly present in social life, open to view and manifest in definite material forms, made actual through a multitude of practices and knowledges, and so entrenched that its apparent invisibility is the product of its sheer ordinariness – it is the very grammar of the 'natural'.

The government of nonhumans

The first concerted attempt to compile official statistics for agriculture in Britain was not made until 1866. The concept of an aggregate economic loss *to the nation* resulting from cattle diseases only became thinkable once agricultural statistics had begun to be established. Prior to 1866 therefore, disease was widely considered an inevitable nuisance to individual farmers, which they had little choice but to accept to a large degree.[10] Indeed, pleuro-pneumonia and foot-and-mouth disease were almost endemic from about 1840, yet no systematic attempt was made to control the spread of contagion. The 1865–66 outbreak of rinderpest or 'cattle plague' and the organizational response it precipitated, however, led to the gradual emergence of a significant new attitude towards cattle diseases. Agricultural statistics made possible for the first time the concept of a definite animal 'population' of a given national territory, to which could be ascribed an aggregate economic value. This meant that for the first time the scale of the financial losses resulting from disease became quantifiable, and the losses of 1866 were quickly condemned as unacceptable.[11] The health of nonhumans suddenly became visibly linked to the economic wealth of the nation, in a connection which would have far-reaching consequences.[12] No longer were diseases of cattle everywhere spoken of as an unfortunate fact of agricultural life; instead they were increasingly decried as an entirely preventable national economic problem capable of being decisively solved by rational organization based on principles drawn from the nascent science of veterinary pathology.

Before 1866 the prevailing view, which had arisen from centuries of practical accommodation to circumstances, was one of fatalistic balance – it held that nature could be manipulated towards human ends only within definite limits and even then only at the cost of contingent re-incursions by nature into what little humans had managed to extract from it; there could be only pyrrhic victories over nature. This now began to be challenged by a distinctly modernist ethos in which nature could be manipulated with precision and its balancing mechanisms thwarted comprehensively by scientific knowledge.

Rather than a result of blind forces of nature, or acts of God, the spread of animal disease came to be seen as a consequence of human practice:

> Until the ravages of cattle plague in 1865–6 led to apprehensions for the safety of the cattle population of this kingdom, the idea of the general adoption of measures for the extinction of disease was never entertained.
>
> (Brown, 1893, p. 283)

> In the period comprised between 1865 and 1893 a complete change has been slowly developed in the agricultural mind in reference to the method of dealing with animal plagues. When, only thirty years ago, the proposition to slaughter in cattle plague was made, it was received with something like a feeling of horror and was dismissed as barbarous. Now a strong condemnation is uttered at the want of energy displayed by a farm which suffers an animal with what was once called the benign foot-and-mouth disease to live. This progress has been slow, and it is not yet complete.
>
> (ibid., p. 286)

This shift was underpinned by an epistemic transformation bound up with the early experimental study of bacteria. No longer could disease be viewed as the product of general atmospheric conditions, of 'draughts' or of heredity, as in what had been the dominant theory of 'spontaneous generation'; instead it was the specific product of a definite organism, which thrived under certain conditions. Herein was an acknowledgement that nature included unseen 'agents' which acted upon the animal body and were the 'true cause' of disease, rather than nature being merely an 'environment' for agents that could only ever be other than nature and thus 'divine', whether in the shape of godly or of human intervention. At the same time as nature was admitted a certain agency, the animal body, its products and its various excretions, became a kind of environment for the invisible and potentially zoonotic agents in nature.[13] The result was to effectively destabilize the established boundaries between subject and object and to radically decentre the agency which was supposed to be the exclusive and defining moment of 'culture'.

> Severe as the trial of the cattle plague was, so long as it lasted, it did more to educate the public mind than all the losses that the country had previously suffered among its flocks and heads, and induced the study of certain animal diseases that the faculty had previously paid little attention to. The diagnosis of diseases is now better understood, as is also their pathology; the theory of spontaneous generation has been upset and the true origin of many diseases discovered.[14]
>
> (Duckham, 1893, p. 281)

Prior to these experimental investigations and the discovery of the bacillus

> by Koch, tuberculosis was supposed to be due to various causes, among which may be mentioned climatic conditions, defective nutrition, exposure to hardships, over-work, excessive secretion – in fact, any and every influence admitted to produce debility. Since Koch's discovery these alleged causes have been considered as only predisposing or preparing the structures of the body for the reception of the disease-producing germs which are the true cause.
>
> (Duguid, 1890, p. 309)

> If anything has been demonstrated beyond all doubt in regard to specific infection it is that in order to induce a specific disease we must first have the seed of it, just as certainly as we require the specific seed to produce a certain plant.
>
> (Brown, 1887, pp. 51–2)

The emergent science of pathology centred upon the question of how to effectively intervene in these newly revealed networks of force or 'actor networks' in order to contain and manage their problematic 'agency'. The development of the experimental study of bacteria meant that subject and object were in danger of dissolving into a myriad of diverse connections between unclassifiable things, shorn of their fixed ontological identities. They had to be re-solidified and repositioned in accordance with the binary categories of humanism; this was the work of the very discourse and practice in terms of which the new knowledge was organized. Scientific pathology was therefore the *technical* struggle of humanist discourse against the *ontological* threat of bacteria. It had to increase the material effectiveness of human action to the point where it could negotiate the blurring of subject/object boundaries and definitions by making itself seem qualitatively distinct from the power of action available to invisible nature. Only in this way could the natural 'agency' revealed by the microscope be made to disappear once again, by being distinguished from a human agency which appeared to transcend it. If only a quantitative difference of degree existed, this had to be stretched materially until it resembled the qualitative difference of kind demanded by 'culture'; human action had to be magnified. The material world was to be restructured until it corresponded to the ontological prism through which it was viewed.

It is through such processes that the very notion of human 'agency' as a preponderant force acting upon a 'nature' outside of itself was made sustainable, by being distinguished from 'natural' forces – which were thereby denied the status of 'agency' – and set on a plane of 'culture' brought into being by this very distinction. Thus the struggle to control cattle disease was far more than an economic issue, for at the same time it was an attempt to maintain a given ontology, which crystallized a specific form of socio-natural order. 'Agency', then, and the subject/object, culture/nature distinction on which it depends, are a discursive effect and not an ontological cause. It is not given in the nature of human practice, but is a historical and material accomplishment brought into being by networks of practice which are themselves neither natural nor cultural in the sense in which these terms are generally understood. Thus 'agency' is the historical effect of the ontological power/knowledge intrinsic to humanist discourse, recognition of which undermines the sociological view of agency as somehow the uniquely given property of human beings. Once this is given, the very coherence of the notion of 'agency' is open to question.

In a conjuncture born of specific developments in statistics and microbiology then, animal disease was de-individuated; it was no longer the chance misfortune of individual farmers, but instead soon came to be regarded as a national crisis demanding collective action.[15] This could not be realised overnight, however; the condition of possibility of the very idea of collective action on animal contagion was the extension of a whole machinery of practices rooted in the newly developing forms of knowledge. Despite the lessons of the 1865–66 cattle plague and the early moves towards centralization and coordination that it precipitated, the government still lacked an administrative apparatus with sufficient resources even to enable the systematic recording of the incidence of cattle diseases.[16] Though only temporarily, for example, the obtaining of official returns for the incidence of foot-and-mouth was abandoned in despair in 1872, revealing the degree to which the foot-and-mouth epizootic was at that time still very much beyond administrative control, and the extent to which the struggle against cattle diseases was a driving-force for organizational modernization.[17] A network of statistical surveillance, regulation and enforcement had to be constructed, not to mention the training and effective coordination of the army of experts who would be needed to apply the new knowledge. This was taken up by the political institutions and became a legislative question; it was at the centre of a raft of parliamentary bills from 1866 onwards, setting out increasingly stringent, progressively standardized measures to be implemented by structures of local authority, as the various layers of local governance became the means to an end that was increasingly national. Indeed, it was the cattle plague above all which gave rise to the Chambers of Agriculture, marking a significant early step towards centralization.

> The necessity of such an organization had long been felt. There were numerous farmers' clubs in different parts of the kingdom besides that in London, yet no effort had been made to bring them into harmony with each other so that their opinions could be ascertained upon any great question, affecting their general interest as a body, until the establishment of the Chambers of Agriculture.
>
> (Duckham, 1893, p. 271)
>
> The Act 53 & 54 Victoria, cap. xiv. July 4, 1890, which was known as the pleuro-pneumonia Act of 1890, transferred the powers of local authorities to slaughter and pay compensation in pleuro-pneumonia to the

> Board of Agriculture, and provided for the payment of compensation out of money voted by Parliament for the purpose.
>
> (Brown, 1893, p. 285)

This process of incremental rationalization and standardization characterized the uneven organizational trajectory of the dairy industry over the next half century. Thus what came to be seen as the national struggle against animal disease was indissolubly linked to an emerging project of social, political and economic centralization.

> The supposed conflicting interests of the 330 borough authorities as against the 92 county authorities were associated with the reluctance which each felt in enforcing the provisions of the Act of 1878, which placed the cost of compensation for slaughter upon the rates for an object that was purely national, and were given as the reason why the Act was never rigidly enforced. It became manifest that pleuro-pneumonia would not be exterminated unless it was taken in hand by a central authority and compensation was provided from the national exchequer.
>
> (Duckham, 1893, p. 277)

Knowledge of the animal body and the control of animal disease then were inextricably a matter of social and political order. The more traditionally political and industrial questions concerning the social organization and regulation of production, and scientific questions concerning the pathology and aetiology of cattle disease, were utterly inseparable in practice:

> Far from being a medical or scientific question, the subject became of as much interest and importance to the stock owner and others connected with the cattle trade, as to the sanitary medical officers and the guardians of the public health.
>
> (Duguid, 1890, p. 305)

> One other important result of legislative supervision has been to drive out of the business the small unscrupulous retailers who lived by fraud, and to throw the great bulk of the trade into the hands of large and responsible companies, such as we see in the Dairy Supply, the Aylesbury, the Express, and others. These cannot be regarded otherwise than as institutions of public health, since they are conducted on lines in strict accord with the principles of sanitary science.
>
> (Wortley Axe, 1892, p. 62)

Already profoundly theological, animal disease was now rendered political, and it was upon this newly constituted political terrain that sanitary discourse and veterinary pathology framed their interventions. Their object of knowledge was the animal body itself, as pathology sought to determine the most effective means to control disease by understanding in detail its precise corporeal constitution. Thus the bodies of nonhumans came to be identified in a discourse and practice that was at once scientific and political as a dangerous source of pollution and disorder demanding urgently to be controlled;[18] the government of nonhumans became at once conceivable and necessary.

> In the past, legislation for the prevention of animal plagues appears to have been adopted exclusively in cases of emergency. Only in modern times do we meet with Acts of Parliament dealing with the whole subject, and conferring powers on administrative bodies to be exercised as occasion may demand.
>
> (Brown, 1893, p. 281)

> The sources of pollution are many and various. They commence with the cow herself.
>
> (Wortley Axe, 1896, p. 51)

Policing the great divide

Only the most general legislation intended to prevent the spread of animal diseases existed before the 1865–66 cattle-plague, and even this was only sporadically enforced. It did, however, focus on the regulation of the movement of animals and animal substances, indicating early recognition of the centrality of movement through space, the circulation of bodies and things, commodities and microbes, to the mode of being of the zoonose itself.[19] The zoonotic plague thrives in dynamic conditions of heterogeneity; its space is the space of flows, the space of mixtures and hybrids, and it was recognized that the movement of cattle to, around, and from markets and fairs was responsible for propagating the disease more widely.[20] This found its nemesis in the fixing, containing, classifying, bounded space increasingly created by sanitary discourse and veterinary pathology throughout the late nineteenth and early twentieth centuries, which sought to re-impose ontological order and security through its categorization, subdivision and analytical ordering of movement. Thus there was a spatial tension between the value of animals as live 'stock', as capital, which depended upon their movement to and from locations of trade and exchange and which therefore involved a fluid use of space, and the necessity of imposing rigid, demarcated space upon diseased cattle in order to inhibit the spread of the disease.

It is telling in this respect that advocates of free trade denounced the measures necessary to eliminate cattle diseases as a form of protectionism; they perceived that combating foot-and-mouth disease, for example, required regulation, inspection, slaughter, and above all the restriction of the movement of 'livestock', not least to and from markets. They rightly intuited that when administrative, bureaucratic power is deployed in the service of a corporeal or socio-political objective – in this case, the elimination of the disease – then the free movement of capital is necessarily impeded. Far from pathology then being indistinguishable as a political technology of humanism from what might be expected to have resulted in any case from the common-sense practicalities of agricultural economics, in fact, there was a significant degree of contradiction between sanitary discourse and organization, on the one hand, and immediate economic pressures, on the other. This was manifest in the protracted struggle within the industry between the advocates of targeted slaughter as a key means of disease control, and those for whom disease was more or less inevitable and the financial loss incurred by slaughter therefore unacceptable:

> Although the evidence given before the several Committees was very conflicting, it tended to educate the country. The evils arising from the numerous local authorities appointed to administer the Acts of 1869 and 1870 were clearly shown. The action of the ninety-two county authorities, who were more conversant with the losses imposed upon the nation by the diseases, was frustrated by the 330 borough authorities, who were induced to look with a jealous eye upon any measure calculated to interfere with the trade of the country, and all who were connected with the foreign cattle trade were eager in their endeavour to show that any interference with their trade would enhance the cost of meat to the consumer. Politicians, too, who had struggled to obtain Free Trade in meat and corn, stoutly maintained that the object was Protection by a side wind, not giving themselves time to consider the evils attendant upon the waste of the home produce, and the necessity of guarding it against such waste.
>
> (Duckham, 1893, p. 274)

Indeed, all attempts to systematically eliminate cattle diseases met with significant and robust opposition, not merely from the borough authorities and from free-traders in Parliament, but from many within the industry who could not see beyond their chagrin at the prospect of short-term losses incurred by the necessity of slaughtering a percentage of their cattle.[21] That this was a matter of fierce contestation in which sanitary discourse and scientific pathology eventually won out by no means weakens the case for arguing that it constituted a critical historical transformation involving an unprecedented realization of humanist discourse and practice within the industry.[22] Far from being universal and monolithic, all discourse is contested and arrayed unevenly in relation to different organizations, social groups and class fractions, particularly during its struggle for hegemony. Humanism is no exception, and must be seen as being borne in differing degrees by different classes, sub-class groups and institutions, at different historical conjunctures. Hence the most obvious sociological reading of this epistemic shift is that it expresses the struggle between different interest groups within the industry, most notably between the urban sanitary authorities and emerging dairy conglomerates, on the one hand, and the smaller, largely rural producers, on the other, a conflict which of course had a significant class dimension. Nevertheless, the fact remains that humanism is irreducible to the human social relations for which it provides a battleground – to think otherwise would be merely to reiterate its anthropocentric assumptions.

In the attempt to manage, organize and thereby eliminate the disorderly and transgressive space of the zoonose, not just the animal body but its immediate 'environment' was broken up into its constituent parts, into the various kinds of matter and forces which might contribute to the transmission of the disease.[23]

> As long as the matters expelled from a tuberculous animal are moist they are not a source of danger so far as infection by inhalation is concerned; but a particle of moisture expelled in the act of coughing may, within a short time after it settles on a wall or woodwork, be so far dried that it will, when disturbed, rise into the air as a dust particle. Everything indicates that this is the way in which the great majority of cases of tuberculosis in cattle are started.
>
> (McFadyean, 1895, p. 658)

Such concerns fed into a widespread debate as to whether it was better to house cattle in sheds for all or part of the year, or to graze them permanently outdoors, and the question of how cattle sheds should be designed in order to optimize sanitation and inhibit the spread of disease.[24] These specific issues were manifestations of a more general discourse on the management of animal bodies and their material conditions of existence:

> The object of the designer certainly should be as nearly as possible to obtain walls and floorings which are impervious to the liquid excreted from the animals which, during a great part of the day in some seasons and still a greater part of the night, are obliged to stay in that somewhat confined space. It is also very desirable to have sufficient ventilation. There is a fiction current that 700 or 800 cubic feet of space is wanted for every cow. If that were a necessity, there is probably not one cowshed in a hundred which could be kept open in a large town. But in my opinion the question of cubic space is unimportant in comparison with the provision of means for the plentiful introduction of fresh air. As for drainage . . . everything should be so arranged that all excreta will either gravitate to the outside, or be easily swept away.
>
> (Brown, 1893, pp. 48–9)

Not only abstract homogeneous space but also what it contained – the precise quantity of air required for ventilation, sometimes even calculated and stated in terms of the minimum cubic volume required per cow; the amount of light, assessed in terms of the number and positioning of windows; and the ease and direction of movement of different fluids, both urine, a source of pollution, which was to be encouraged to drain away, and milk, which was to be kept free of all pollution – these became the objects of a cumulative instrumental knowledge of material interactions.[25] The formulas used were somewhat arbitrary and inconsistent at first, varying considerably between local authorities; they were derived ad hoc from the design of army barracks and boarding schools, where similar questions of corporeal government and disciplinary organization had recently begun to be systematically addressed in relation to human beings.[26]

> Few reliable tests have been made on animals, but on man we have sufficient data to guide us in the present enquiry. Army barracks are designed on a basis of an allowance of 600 cubic feet per man; and, calculating the weight of an average Ayrshire cow at 8¼ cwt, and that of a man at 11 stone, the one will be exactly six times that of the other . . . If in a byre the air is to maintained as pure as it is in our Army barracks, and the same provision is made in both cases for ventilation, then the cubic air space necessary for an Ayrshire cow, according to this basis, will be 600 ft. × 6 = 3,600 cubic feet. This is a cubic space six or seven times greater than the average cow is allowed, so that either the soldier gets too much space, or the cow too little, or else the basis of calculation is altogether wrong. Besides Army barracks, we have, however, other instances where the Government stipulate for a certain cubic space for each human being. A well known one is our Board Schools. In this case 10 square feet of floor space is demanded for each child.
>
> (Speir, 1896, p. 8)

> The principal point to be attended to in keeping the animals clean and comfortable is to have the stall of a size suited to the size of the cow. This is attained by having stalls of different dimensions in different parts of the byre, and some farmers manage this by having one lot of stalls of one size and one lot of another; but in my own case I prefer to have then graduating from the largest to the smallest, i.e, the kerb or edge of the bed is, say, 8ft. from the front stall or partition at one end of a row of cows, and 7ft. at the other, so that the stalls graduate from one to the other. In this way the cows are sized out in regular order, and the perspective and general appearance of a byre is greatly improved when they stand in order, thus, like a regiment of soldiers.
>
> (McConnell, 1899, p. 85)

In part, this was driven by economic exigencies, as its financial benefits gradually came to be more widely acknowledged, but it also had an ontological politics of its own, meticulously policing the boundary between 'culture' and 'nature' and in the process bringing it into being, not just discursively but materially, and with very material effects:

> Since the passing of the Local Government Act of 1889, and the Public Health Acts, 1867, 1870, and 1878, with the Orders in Council relating thereto, sanitary matters pertaining to byres, stables, etc. have been enforced in a manner previously unknown.
>
> (Speir, 1899, p. 7)

> A few years ago there was no Act controlling the sanitation of dairies and cowsheds. Today an army of sanitary inspectors periodically visit the dairies and cow-sheds to ensure their being in a proper sanitary condition.
>
> (Pocock, 1898, p. 19)

The first analytical tool of this sanitary practice, by means of which it sought to break down the animal body and its environment into discrete functional parts, was an internal/external dichotomy. This allowed an essentially Cartesian distinction to be made, at once ontological and strategic, between what was inside the animal and what was outside, facilitating a careful mapping of the various thresholds of the body which might carry the infectious bacteria from the animal interior to the human exterior, which is to say from 'nature' to 'culture'. Thus the potentially infected animal body became a material problem of governmentality conceived in terms of pollution and containment – how to manage the body's processes in such a way as to restrict the spread of what it 'contained' to the 'outside':

> It will be convenient if I contemplate the subject in two distinct aspects – milk, as susceptible of contamination from within, from the animal's own system; and then as a fluid liable to be contaminated by contact with certain noxious matters which exist outside the animal's body.
>
> (Brown, 1887, p. 44)

> Most of these disease germs come from outside contamination; but there is one class which we have every reason to believe is very commonly found in milk which comes from the cow itself, viz. the tubercle bacillus – that which produces tuberculosis or consumption.
>
> (Aikman, 1896, p. 45)

> Besides the products of the disease which spring out of the cow herself, others derived from extraneous sources and equally potent for evil may, by attaching themselves to her in an accidental way, ultimately reach the milk.
>
> (Wortley Axe, 1896, p. 51)

> There are various ways in which the bacilli bred in the interior of an animal may gain access to the outer world.
>
> (McFadyean, 1895, p. 657)

This problem was posed most starkly in relation to dairying, which was then beginning a long period of steady expansion, for milk is easily read as the product of an animal's interior which is systematically extracted in order to be consumed externally. The implications were not lost on contemporary agricultural writers:

> It is possible, but by no means proved, that the infective matter was conveyed through the agency of the milk.
>
> (Brown, 1887, p. 52)

> It is within the knowledge of all our readers that milk not infrequently acts as a propagator of disease by being the medium for carrying certain pathogenic or disease-producing germs.
>
> (Aikman, 1896, p. 47)

> It is obvious that tubercle bacilli passed out in any of these ways may easily gain access to the body of another animal, and thus start a new case of the disease. Thus, milk coming from a tuberculous udder is certain to infect the calf that consumes it.
>
> (McFadyean, p. 657)

> The re-introduction of foot-and-mouth disease into our dairy herds adds another to the list of contagious disorders communicable to man through the medium of milk.
>
> (Wortley Axe, 1892, p. 64)

> There exists no doubt in the minds of the most advanced pathologists that human tuberculosis is largely due to specifically infected milk.
>
> (ibid., p. 65)

> One of the most serious, and at the same time the most prevalent, diseases of dairy cattle which threaten human life through the agency of milk is that fatal affection tuberculosis, or, as it is commonly termed, consumption.
>
> (ibid.)

Meat passes through a critical transformation before being consumed by humans; in the practice of cooking it is separated from the animal body, the traces of nature are effaced before its consumption, and in the process it becomes not flesh but 'meat', and part of 'culture'.[27] Thus cooking is a humanist practice, serving to 'civilize' the animal body and complete its transformation into an object of human consumption, which began with its death. Milk, on the other hand, which is necessarily extracted from a still living animal, in the mid-nineteenth century underwent as yet no such 'civilizing process', and consequently posed a danger to human 'culture' at once material and ontological.

> Meat is always cooked before being used as human food, and it has been demonstrated by Koch, Lingard, and others that a boiling temperature for a few minutes is sufficient to destroy the vitality and infective power of the bacillus; but milk is generally used as food without cooking, more especially in the case of children. Tuberculous milk must therefore be looked upon as dangerous and likely to be the means of producing the disease in young or weakly subjects consuming it.
>
> (Duguid, 1890, p. 314)

> The results of the feeding experiments all tend to prove that the milk from tuberculous cows, if given to animals in the uncooked state, possesses a very much greater infective power than the flesh.
>
> (ibid.)

> It is true that the deposits of tubercle are seldom found in the meat, and the feeding experiments have proved that even in the advanced stages of the malady, when the milk proved infective to guinea pigs and rabbits, the flesh in the raw state was harmless.
>
> (ibid.)

Now both the scale of the liquid milk trade and its importance within British dairy agriculture increased significantly in the last decades of the nineteenth century.[28] This was facilitated by the expansion of the railways and driven by the pressure of foreign competition in the less perishable dairy products, which was brought to bear by refrigeration technology. Against this background, tuberculosis – the zoonose most readily transmissible through milk – became the problematic object for a network of disciplines embodying various specialized knowledges, both socio-political and medical-scientific. These disciplines laboured to impose controls on the passage from internal to external, to purify it, render it amenable to human control, and subject it to the anthropocentric categories it was so apt to undermine; in this way they were integral to the technical operationalization of humanist discourse at the heart of modernization. This project was enormous in scope, as complex and fragmented as it was ambitious, and it is unsurprising that the aetiology of tuberculosis was still widely contested even at the turn of the century.[29]

> During the past eight to ten years tuberculosis, or consumption, has received more attention than any other disease of either man or animals. The reason for this has been not only that the belief in its contagious nature was becoming more generally accepted, but the discovery by experiment that the disease in man and animals was the same, and *intercommunicable* from one to the other. When this discovery was made, the

> question naturally arose how far the flesh and milk of tuberculous animals might prove injurious to man when used as food.
>
> (Duguid, 1890, p. 305)

With this in mind, the full significance of the early stages of pasteurization and of the regular tuberculin testing of dairy herds in the last years of the century can begin to be appreciated. As indeed can the importance of the whole ensemble of emergent practices and technologies associated with the drive for 'clean milk' in the early twentieth century, through which the blurred and complex spectrum of continuous material relations between animals and humans were gradually fashioned into a clear-cut dualism.[30] These included innovations in transportation and delivery which were designed to reduce the risk of contamination, leading to the earliest moves towards the doorstep delivery of bottled milk, in a further technological effacement of the traces of its production;[31] they also included the intensification of long-standing efforts to produce an efficient milking machine as the cornerstone of a strategy of increasing mediation between milker and cow, a process which would eventually, decades later, see the first fully mechanized dairies in which both cow and milk were untouched by human hands. Thus practices and relations which endlessly criss-crossed 'nature' and 'culture' were progressively reshaped until they were brought into line with the conceptual certainties of humanism, in a process which, once completed, did indeed provide a material and practical basis for humanist discourse; by such circular processes modern 'culture' and the notion of 'the human' intrinsic to it were made materially and historically actual.

Civilizing the animal body

With the growing confidence of analysis by dissection, and the accumulating knowledge of microscopic, hence 'invisible' organic processes, the inside of the animal body, the ontological space of 'nature', of potential pollution and disorder, became equated epistemologically with 'truth'. Whereas the outside, the space of 'culture', was relegated to the status of mere appearance, which may or may not reveal the true inner condition of the animal:

> A beast in good condition fit for the butcher, and *apparently* in perfect health, may on slaughter be found to have tuberculous deposits in its lungs and other *internal* organs, while on the other hand, the 'waster' or 'piner' in the most emaciated condition may, when killed, be found quite free from tubercle or other organic disease, the emaciation being due to other general causes of debility, such as excessive secretion of milk or insufficient quantity or quality of food.
>
> (Duguid, 1890, p. 306)

Inner 'truth' was made synonymous with science, with concrete knowledge of nature, and with expert or specialist knowledge, whereas outward appearance was associated with the practical and everyday knowledge of the small farmer. Hence the acquisition of scientific knowledge was inscribed as a passage from external to internal, from culture to nature; science was to penetrate into the animal nature in order to colonize it for culture and sanitize the process of its externalization for human consumption. This is also detectable in what were then nascent scientific nutrition and breeding practices; the transition from breeding for outward appearance towards breeding for functional efficiency, for example, involved the same epistemological movement from external appearance to internal 'truth'.[32] It is not without significance that in both configurations the nonhuman being itself fails to signify. Where the animal is identified with its outward appearance, it is plausible to reject the notion of animal 'consciousness' or 'mind', which humanist metaphysics is located vaguely 'inside' and yet not *of* the body – it is an incorporeal substance; thus the animal becomes a simple observable object, which by definition is without 'mind'. Conversely, if the animal is identified with its 'inside', and this turns out to be reducible to a set of processes amenable to scientific analysis and divisible into functional parts, then similarly the whole animal itself apparently disappears – its 'soul' is expunged. In its most extreme expression this led to the view that nonhuman animals were nothing but organic machines, like clockwork of flesh and blood. It is therefore necessary to see the two sides of the internal/external distinction as different faces of the same regime of truth, as an analytical extension of the Cartesian mind/body, subject/object distinction, and therefore as a mechanism of humanist discourse.

Upon this distinction between internal and external, the discourse of pathology was articulated as a set of knowledge claims with an implicit hierarchy:

> In describing tuberculosis it becomes necessary to decide what it really is, and how it is to be distinguished from other diseases which resemble it. As already stated, any deposits in the form of little lumps or nodules in the tissues or organs of animals were formerly described as tuberculous; but since the true nature of the affection has been made out, a definite meaning has been attached to the term tubercle by the scientific man, though to the practical man employed in the slaughter house, conversant only with the naked-eye appearances, no such distinction is evident. It is therefore necessary to regard tubercle from these two points of view: the first, the practical every-day aspect which meets the eye of the slaughterman or ordinary observer; and the second, the view which the pathologist and bacteriologist takes of it.
>
> (ibid., pp. 307–8)

> Looking at tuberculosis from this practical and every-day aspect, there is often great difficulty in recognising its existence in the living animal, or even determining by the unaided eye its presence in the tissues or organs

> after death. In dealing with tuberculosis as a disease of animals, or with the question of the fitness or unfitness of the flesh or milk of tuberculous animals for human food, the practical view must always come first, and the experimental or bacteriological view follow as a means of confirming or correcting the evidence obtained from symptoms during life and appearances after death.
>
> (ibid.)

> The microscope has revealed more numerous secrets as to the germs of diseases which could not have been brought to light in any other way.
>
> (Duckham, 1893, p. 281)

This was not only a hierarchy of knowledges, but a hierarchy cast in terms of distinct visibilities, ways of knowing-by-seeing, with naked-eye visibility requiring confirmation or correction from the expert microscopic visibility of the pathologist. These visibilities were intrinsically political, not only because they were unevenly distributed between social groups, affording legitimacy to the urban local authorities at the expense of rural producers, but because they were the technical means by which the management of nonhuman bodies in space was achieved, both at the level of the minutest anatomical detail, with the dissection and mapping of the body's processes in relation to its environment, and at the level of the utmost statistical generality, with the numerical mapping of the animal population and its local and regional distribution. By knowing animal bodies in space at these contrasting levels of abstraction it was possible to organize their movement in order to produce a tightly regulated, panoptical, binary space, which militated against the disorder of the zoonose; through such methods humanism was spatially entrenched.

The hierarchy was also a temporal one, being associated with the passage of the animal from life into death; thus symptoms in life become appearances in death. The one is suffered by a subject, the other observed in an object, though prior to dissection and microscopic examination both have the status of the everyday or practical rather than the scientific, and might therefore prove misleading. Of course death itself is a double-edged sword for humanism. The corpse in corporeality, death is an existential taboo against which all humanism strives; the penultimate conclusion of disease, it is also the key means of its destruction. Thus systematic slaughter was always the blunt instrument at the heart of disease control, even while the forms of knowledge and networks of practice through which this shock therapy was administered became increasingly precise.[33]

The modernist belief that 'nature' could be made subservient to 'culture', natural 'agency' subsumed under human 'agency', and natural forces controlled so completely as to render them passive objects to the human subject, was paralleled by a shift away from the assumption that the per capita consumption of milk was essentially inelastic, towards the view that it was highly if not infinitely malleable. Hence the organized and systematic campaign to eliminate cattle diseases preceded a methodologically similar campaign to raise levels of milk consumption more rapidly than the increases attendant upon population growth. A prerequisite for this was the perceptual transformation of milk from its popular image as a dirty, potentially infected and adulterated fluid to that of a clean, pure and nutritious staple food. In part, this objective was pursued through legislative measures against adulteration, but it also meant tackling milk-borne tuberculosis.[34] Thus the project of veterinary sanitation had to be extended from its ongoing battle with cattle-plague, foot-and-mouth disease and pleuro-pneumonia, to encompass regulation of the production, transportation and consumption of liquid milk, as this became increasingly pivotal in the campaign against cattle diseases.

Tuberculosis was nevertheless the last of the major cattle diseases to be named in legislation, reflecting the fact that right until the last decade of the nineteenth century, and even into the twentieth, despite the warnings of some veterinary pathologists, there was considerable popular scepticism towards its contagiousness, and particularly towards its communicability from animals to humans.

> The question has often been asked, how far and to what extent is tuberculosis contagious? Even at the present time, although the disease has been produced experimentally by inoculation, inhalation, and ingestion, many people do not *believe* in its contagious nature – at least not in the ordinary sense.
>
> (Duguid, 1890, p. 313)

> Many who admit that a man may catch small-pox, or scarlet fever, or an animal may take foot-and-mouth disease, from having been in contact with patients suffering from those affections, would laugh at the idea of either men or animals catching consumption.
>
> (ibid.)

> I am of the opinion that the danger of consumption being conveyed by milk from the bovine to the human being has been greatly exaggerated.
>
> (Horner, 1909, p. 36)

> It is not the object of this paper to discuss the question as to whether the consumption of milk can or cannot produce tuberculosis in human beings. The experts are not yet fully agreed upon this point. That being so, it is incumbent on mere laymen like myself to stand aside while the experts fight this battle out among themselves, and arrive – as no doubt they will in due course – at some definite conclusion.
>
> (Sadler, 1909, p. 68)

Indeed, for some considerable time, the terms used to discuss the contagiousness of tuberculosis tended to focus upon either its transmission from human to human, or to a lesser extent upon its transmission from animal to animal, without giving any significant consideration to trans-species communication from animal to human.[35] This was symptomatic of a powerful anxiety surrounding zoonotic disease more generally. Thus, even in the 1890s, when it was becoming impossible in the face of mounting evidence – particularly the soaring rates of urban infant mortality associated with the trend towards substitution of dairy for breast milk – to avoid considering the possibility of human contraction of tuberculosis by consumption of animal substances, this was dealt with as a question of 'public health', allowing the material connection with the dynamics of the disease among the animal population to be carefully avoided.[36]

> Legislation for tuberculosis must be considered from two different points of view:
>
> 1. In the interests of the public health, restricting or prohibiting the use of tuberculous meat and milk as human food. 2. Dealing with it as a contagious disease of animals, and adopting suppressive measures to eradicate it from the country.
>
> (Duguid, 1890, p. 318)

> Animal health and wholesome milk are the two chief objects of dairy sanitation. Although the one does not necessarily imply the other, it is nevertheless of the first importance to uphold it, both for sanitary and economic reasons.
>
> (Wortley Axe, 1896, p. 51)

Thus a disease which transgressed the human–animal divide, a natural 'agent' which threatened to undermine the coherence of 'culture' by polluting its anthropocentric domain, was artificially separated into two separate questions: the one human and concerned with social and political questions of public health, the other animal and concerned with purely technical questions of effective suppression; on the one side were positioned consuming human subjects, and therefore culture and sociality, while on the other were only nonhuman objects and therefore nature. Onto this dualism was mapped a further division between society and economy, with public health deemed a social problem and animal health an economic problem centring upon the damage to agricultural productivity.[37] The measures necessary to eliminate tuberculosis from milk intended for human consumption were therefore considered quite apart from the question of how to eradicate the disease among animals.[38]

This bifurcation of the problem led to the later tensions between tuberculin testing and pasteurization. Different principles were at work here, and the relation between them was the cause of some dispute: Why was pasteurization necessary if tuberculin testing could assure healthy cows? And why undertake tuberculin testing if pasteurization provided certain sanitation of the milk? This debate reflected two very different ways of thinking of the sanitary project and the campaign for clean milk. Pasteurization involved the 'cooking' of the milk and hence was resisted by some critics as a denaturalization of the product. Tuberculin testing, on the other hand, was far less reliable a safeguard, but it represented a different philosophical approach entirely, tackling the disease 'at source' among the animal population, so that 'healthy' animals could produce 'natural' milk. In this way the tension between pasteurization and tuberculin testing was structured by the humanist dislocation of animal health and public health. Though a product of the animal body, milk became an object of 'culture' in the process of its sanitization and thus – as part of 'culture' – regained a kind of subject status in relation to the animal body which, though living, was in turn inscribed as an inert 'nature' outside of 'culture' and therefore an object, albeit a complex and organic one. Even as sanitary practices then began to reinforce the boundaries of the human domain against the zoonotic disorder materially, sanitary discourse was organizing scientific knowledge of the disease according to a division between human and nonhuman.

Conclusion: interpellating the human

> If we belong to the human species, we do so not because we have any particular qualities, but because that is how we draw the boundaries of our lineage.
>
> (Fernandez-Armesto, 2004, p. 4)

The three to four decades following 1866 in British dairy farming saw the development of a nexus of practices and knowledges in terms of which the bovine animal was discursively positioned and ontologically classified in contradistinction to the human. Their effect was to purify the boundary between the human and the animal, crucially by rendering the animal body – and particularly liquid milk – 'safe' for human consumption. This accomplished far more than its most obviously practical purpose – preventing the zoonotic communication of tuberculosis to humans, for in the process it contributed, in concert with innumerable similar mechanisms operating ceaselessly, to the constitution of the very sphere of 'culture'; that powerful myth at the heart of modernity of a uniquely human realm, above and beyond 'nature', a sanctified domain, apparently given, but in fact perpetually carved out in discourse and practice. By literally disinfecting the problematic material relation between human and nonhuman, the anthropocentric polarity was made possible between 'nature' – the animal body or raw material – and 'culture' – its consumption by humans. A similar process could be observed with pasteurization, in which the material flow from an earthly animal nature to a transcendent human culture was subjected to precise sanitary regulation and thereby cleansed of its intrinsic heresy. Technical control at the level of material details was the means of ontological control at the level of metaphysical abstractions. Like the Inquisition, the faith was defended by a whole machinery of bodily manipulations. Milk was thereby 'civilized' and rendered 'clean'; where there once had been a deeply problematic animal substance and a bearer of zoonotic agency, now there was a purified cultural product, ripe for the mass marketing of the twentieth century, and no longer so pregnant with questions concerning our real relations with the other animals. This was above all a material accomplishment requiring constant vigilance and ceaseless labour for its continuing plausibility; it had to appear at all times to be the natural order of things.

The development and dissemination of sanitary discourse and veterinary pathology were intimately connected to this practical and epistemological operationalization of humanist discourse; only by regulating and controlling animal bodies and animal diseases with precision could their intrinsically problematic relationship to the category of the 'human' at the heart of modernity be suppressed and managed. 'Animals' are not a self-evident category but have to be historically defined and constituted by an array of discourse and practice. Administrative and organizational rationalization was critical in the general diffusion and co-ordination of such practices, pointing towards the intimate relationship between humanist discourse with its ontological separation of humans from nonhumans and culture from nature, on the one hand, and modern governmentality with its construction of a public sphere, its distinctive technologies of rule, its particular political ways of knowing, on the other. Thus the humanist practices underpinning the constitution of a sphere of 'culture' were articulated within distinctly modern forms of social and political organization, which were themselves, in turn, dependent for their rationale upon the historical category of the 'human' thereby produced. The internal connection was the resulting notion of the human individual as a cultural subject, a free and rational agent separate from nature – the secular soul of modern politics. In this way the discourse of humanism is deeply political; it is inseparable from the epistemological conditions of possibility of modern forms of sociality and power. Its effect is not directly to control human subjects, though it certainly facilitates such control, but more importantly to classify and thus produce them as 'human'. By organizing the historically specific ways in which it is possible to know oneself as 'human' vis à vis the knowable 'animal', humanism produces that definitive category without which 'culture' itself would be unthinkable.

4 Purifying milk

Knowledge, sanitation and discipline

> The medical supervision of diseases and contagions is inseparable from a whole series of other controls.
>
> (Foucault, 1991, p. 144)

Introduction: the problem of 'agency' in humanist discourse

> Social science is always founded upon a categorical distinction between the ideality of human intentions and purposes and the object world upon which these work, and which in turn may affect them. There is little room to examine the ways they emerge together in a variety of combinations, or how so-called human agency draws its force by attempting to divert or attach itself to other kinds of energy or logic.
>
> (Mitchell, 2002, p. 29)

The previous chapter focussed predominantly upon the transformations of dairying and the emergence of the modern dairy industry in the quarter-century following the cattle plague of 1865. It also introduced an argument concerning the ontological significance of tuberculosis as a zoonotic disease transmissible through dairy milk. This chapter picks up these threads and deepens some of these arguments by developing a more localized analysis of the development of milk testing and inspection, and by looking more closely at the struggle against milk-borne tuberculosis which took place during the period of 1890–1920. In terms of its historical argument, this chapter posits a shift in the rationale of sanitary discourse and practice and suggests that this was bound up with a wider shift in the mode of realization of humanist discourse. I identify the city as critical in this shift, particularly in terms of a changing spatial imaginary of the city in relation to the country, which was inscribed within urban sanitary discourse and organization. In my understanding of the changes which were occurring in the city at this time I am indebted to Patrick Joyce's analysis of the transition from 'the sanitary city' to 'the social city' in the late nineteenth century. He argues that there was a shift from a medicalized view of the city to a social, and indeed a cultural, one:

> The social in this view of the city became *sui generis*, something with its own laws of what was called 'society', a particular sort of reality to be known by social science. The 'social' was now itself to do the explanatory work. Certain things came to be known to this science as 'social' things, and therefore different from anything else.
>
> (Joyce, 2003, p. 172)

My argument is parallel but somewhat different. I believe that 'the social' already infused the earlier 'sanitary city', for this medicalized view of the city was always at the same time a social one; it merely viewed the social in sanitary terms, as is clear from its tendency to equate social disorder with disease and disease with social disorder. What Joyce identifies as the emergence of the 'social city' was actually the emergence of a new kind of 'social', that is, a 'social' constituted by different means. To put it another way, the emerging 'social city' was the product of a modulation in the ontological politics through which the category of the 'social' is perpetually inscribed upon certain phenomena, in its 'inscription devices'. Recall my argument that the enactment of 'the social' depends upon operationalizing the humanist principle that there is a domain of 'the real' consisting of meaningful relations between human subjects, a domain of exclusively human subject–subject relations. Hence the 'social' is forged within epistemic and material practices in which human 'subjects' are constituted by being differentiated from nonhuman 'objects'. From this point of view, the shift identified by Joyce as a shift from the 'sanitary city' to the 'social city' was actually a shift in the modality of modernity, in the mode of actualization of its ontological structure, a methodological shift in which the 'social' in fact *ceased* to be *sui generis* for the practices constitutive of it, as a condition of its reproduction under changing material and epistemic circumstances. This was precisely what enabled 'the social' to continue to be *experienced* as *sui generis*. It can also be understood in terms of a change in the dominant mode of ontological ordering.

The 'social' in the modern sociological sense does not pre-exist this shift, but emerges from it. This 'social' is not an ontologically real object or domain of some kind which can be identified (tautologically) as the ground of all 'social' explanation. Indeed, any attempt to project 'social' explanation onto earlier phenomena is really just an affirmation of this form of explanation, as distinct from an engagement with the processes that gave rise to the category; it is an exercise in retrospective ontological realism. The alternative is genealogy, an approach to history which tries to account for the constitution of knowledges, discourses and domains of objects, without grounding its explanation in a transcendental subject or indeed a domain of relations between such subjects, that is, in an ontologically prior 'social'. The ontological question cannot be delayed indefinitely though. To be sure, it is always already epistemological, but it would be vain to insist that this frees us entirely from the problem of ontology. For though it may only ever be knowable through the epistemology which shapes it, some residual ontology is always implied.

It is therefore necessary to posit a residual and fluid ontology consistent with the implications of this position; anything less is disingenuous. Such an ontology would acknowledge the corollary of the genealogical approach, which is that the 'social' emerges from something else, something more-than-social or other-than-social, something materially heterogeneous in a way that the concept of the 'social' fails to capture. Given my general argument concerning the role of humanist discourse in producing the 'social' as a mode of organization of experience through processes of human–nonhuman separation, I contend that this more-than-social consists of heterogeneous networks or assemblages of human–nonhuman relations, overlapping and interwoven and constitutively mixed in hybrid and contingent formations. In this view I am broadly in line with actor network theory. As I have argued though, human–animal relations occupy a critical position in this, principally because the specific inter-corporeality of these interspecies relations and encounters makes them uniquely problematic for humanist, dualist ontology; thus they tend towards moments of presence and ambiguity which must be contained and sanitized by humanist technologies. Against this background the aim of this chapter is to deepen my analysis of the actual operation of the humanist ordering processes which subdivide these problematic relations into compartmentalized 'social' and 'natural' phenomena.

I have already suggested that humanist discourse is instantiated within and through diverse rationales and motivations, and that the late nineteenth century saw a restructuring of the mode of operationalization of humanist discourse, of its political technologies. In this chapter I give more substance to these ideas by arguing that humanism increasingly became infused within structures of expert technical knowledge and administrative practice, so that the growth of expert authority vis-à-vis everyday practical knowledge in the late nineteenth and early twentieth centuries involved the imposition of a new ontological politics of human–nonhuman relations, constituted through myriad technico-political practices. This involves a more detailed consideration than I have hitherto attempted of the quasi-agents of humanist processes and of how humanist discourse is enacted within their material and epistemic practices. But if 'agency' emerges from such humanist processes rather than preceding them, then any asymmetrical focus upon social 'agents' must be qualified by acknowledging that such 'agency' is at least partly an explanatory fiction, for it is always inseparable from its enactment within complex and contradictory relationships with nonhuman entities. As John Law has argued, such entities are not themselves 'agents' in any normal sense, but they nevertheless 'act', in that they 'make a difference'.[1]

Like 'the social', therefore, 'agency' does not exist as such; it only comes into being post hoc and by attribution. The practices in which these attributions are made are humanist discourse-practices, which constitute agency as social and human precisely by othering the nonhuman. Once othered, these nonhumans which nevertheless 'make a difference' are able to be incorporated into networks of humanist discourse-practice as 'objects' of human agency, but occasionally their effect is to undermine humanist processes, to disrupt humanist discourse and to destabilize the dualist architecture of 'culture'. In such cases the recalcitrant nonhumans in question must be technically and discursively managed in order to contain their problematization of humanist ontology. In the previous chapter I identified zoonotic diseases as an important case of such ontological disordering, and described tuberculosis as an entity materially erosive of the integrity of the 'social' in its agnosticism and its flows of intercorporeality. Pursuing this argument in more depth, much closer attention is given in this chapter to the technico-epistemological and administrative restructuring of humanist discourse in response to zoonotic tuberculosis transmissible through dairy milk, which I explore in terms of two closely interconnected historical developments:

The first and most important of these was a shift in urban sanitary organization, precipitated by the emergence of a new epistemic object, a new nonhuman 'agent', namely zoonotic tuberculosis conveyed through milk and knowable by means of the microscope. I posit a transition, gradual but fundamental, from a Victorian sanitary paradigm in which urban 'public health' sanitation and 'agricultural' disease prevention – and by extension the 'social' and the 'natural' – were utterly dislocated, conceptually, organizationally and practically, to a modern 'purifying' paradigm in which the commensurability of these domains was practically acknowledged in certain kinds of material and knowledge-practices. But this commensurability was only acknowledged in order to perpetually delete it, or rather to 'cross it out' without erasing it completely, thus rendering it an absent presence. This is significant because it marks the moment in which a new anatomy of the 'social' began to be constituted. This 'social' was an acutely technical accomplishment, engendered through myriad processes of separation of the human from the nonhuman, which were increasingly technoscientific in nature and micro-managed by 'experts'.

In making this argument I draw upon Timothy Mitchell's analysis of the pivotal role of expert knowledge and 'techno-politics' in the formation of the economic as a distinct sphere in modern Egypt. Fascinatingly, Mitchell identifies plagues of mosquitoes in their role as vectors of malaria as critical 'agents' in the networks which brought the modern 'economy' into being.[2] In my account, the problematic 'agency' of milk and tubercle bacilli parallels that of mosquitoes and malaria in Mitchell's analysis, being critical factors in the constitution of the version of 'the social' enacted within the new sanitary paradigm. I argue that the microscope rendered knowable a nonhuman agent profoundly disruptive of the Victorian model of sanitary organization in its transgression of the human/nonhuman divide, which underpinned the dislocation of urban public health from agricultural disease prevention. The resulting ontological insecurity fuelled decades of epistemic struggle over the communicability to humans or otherwise of bovine tuberculosis, in which the knowledge and authority of the ascendant class of experts and their power to define and constitute the social by knowing, classifying and regulating its boundaries with the nonhuman were progressively established.

This process was circular, for it was the technical expertise embodied in the microscope which had first called forth 'tuberculosis' as an epistemic object and thereby disordered the Victorian paradigm. Ironic then that it was also technical expertise, in the form of sanitary science and microbiology, which purveyed the means to re-impose ontological order and re-establish humanist categories, by organizing systematic milk testing linked to farm inspection, thereby extending the disciplinary technologies of 'culture' into 'nature' and containing the disorder within the natural. Thus the experts were able to cement their own authority by providing a technical solution to a problem which they had themselves been instrumental in creating. I argue that this was not just a shift in technical and administrative practice, but at the same time in ontological organization, that it marked the emergence of an apparatus of humanist disciplinary practices intrinsic to the constitution of the modern 'social' of the twentieth century. Clearly it is not possible to adequately explicate such a transformation while remaining at a high level of geographic generalization. Fortunately this is not necessary, for I am in agreement with Patrick Joyce and Simon Gunn that Manchester was not only broadly representative of change in Britain as a whole, but in many ways ahead of and initiating such change.[3] I therefore begin by focussing upon the transformation within Manchester in some detail, so that it effectively forms an extended local example or case study within a case study.[4] The second major development explored in this chapter is the growing authority of urban sanitary experts vis-à-vis the traditional knowledge of dairy farmers. I examine this in terms of the epistemological politics of the contested knowledge itself, specifically the prolonged debate as to whether tuberculosis of cattle and tuberculosis of humans were in fact the same disease and whether it was transmissible from one species to the other. The twists and turns of this struggle are illuminating not just for the ways in which humanist discourse was realized within constellations of competing interests and knowledge claims, but also the ways in which these shifting networks were variously enabled, disrupted or transformed by the materially heterogeneous connections between nonhumans of diverse kinds, not least tubercular bacilli, nonhuman animals, liquid milk and the microscope.

This chapter outlines a series of complex embroilments then: the embroilment of humanist discourse and sanitary organization; the different embroilments of the city and the country in the spatial imaginary of sanitary discourse; the embroilment of technical knowledge-practices and late modernity; the embroilment of the microscope and tubercle bacilli; the embroilment of the laboratory and the farm. These embroilments were not causal or sequential relationships; one can identify in them fragments of something approximating 'causality', but an account which restricted itself to piecing together these causal fragments in the name of methodological rigour would merely serve to reinforce the security of this model of explanation, this kind of residual positivism. It would do little to illuminate the kind of parallels, asymmetries, deletions, absences, mediations and emergences which can be glimpsed in these relations but are in no sense causally or even empirically 'present' as such, and which thus cannot be established in any decisive way against other interpretations.[5] In the end, therefore, I think it worth abandoning the epistemological security of a common-sense realism in order to pursue the possibility of saying something new. Although this does not preclude an approach which is attentive to historical details, it does refuse to be circumscribed by realist notions of what can be done with these details. That is my understanding of how best to apply the spirit of the sociological *imagination* within a symmetrical approach to heterogeneous relations. For this reason, rather than artificially separating the entanglements described in this chapter and arranging them into a linear or causal narrative for heuristic purposes, they are to a certain extent interwoven and overlapping in my account, as they were in the field of events.

Power of experts: the techno-politics of humanism

> The Inspectors under the Food and Drugs Act, and Medical Officers of Health in urban districts – a disciplined army of them – are on the war path, looking out for delinquent dairy farmers and dishonest milk sellers everywhere, leaving no stone unturned.
>
> (Sheldon, 1909, p. 27)

> The microbes seemed too arbitrary in their attacks on humans. When cleanliness had meant health and dirt meant disease, the science of hygiene had taught morals; but the lesson weakened the more that *anyone* was at risk of bacterial invasion – as often seemed the case when late nineteenth century medical scientists tried to distance themselves from older styles of public health.
>
> (Pickstone, 2000, p. 50)

The last decade of the nineteenth century witnessed a growing anxiety over tuberculosis. Newspapers began to make the link between infant mortality and infected milk on a regular basis, reinforcing a pervasive identification of milk with the disease.[6] In particular, the belief took hold that dairy milk from the country was responsible for a very significant proportion of the infant mortality from tuberculosis in the towns and cities.[7] It was this climate of widespread public suspicion of dairy products, and especially of liquid milk, which facilitated the development of the sanitary campaign against milk-borne tuberculosis.[8] Though economic factors also became a driving force for such sanitary reform, they could not have driven anything without the political, administrative and discursive emergence of 'clean milk' as a public service, moral duty and national philanthropic 'good'. The establishment of this framework shaped economic practice by transforming what was profitable and what was not; thus economic exigency followed rather than determined political and organizational shifts which effectively defined the contours of economic rationality. Indeed, the process of concentration of ownership and the emergence of large dairy conglomerates were preceded fully ten years by the beginnings of the development of substantial sanitary regulation of the milk supply. This took the form of a growing apparatus of milk inspectors, medical officers of health and local government sanitary officials, engaged in the enforcement of orders enacted under legislative authority conferred by the Board of Agriculture and the Local Government Board.[9]

The struggle against tuberculosis provided the legitimation for the construction of this disciplinary apparatus. This in turn brought to a head the tension between rural and urban interests, and particularly between metropolitan veterinary inspectors and medical officers of health, on the one hand, and dairy farmers and producers' associations, on the other. These groups posited sharply opposing accounts of contamination, accounts which inscribed conflicting moral and sanitary geographies upon the imagined relations between city and country, and which vied with each other in the larger struggle between experts and producers for moral authority.[10] Dairy farmers typically suggested that it had always been urban milk sellers who were responsible for unclean and adulterated milk, and that the role of milk-borne infection and bacteria in producing high rates of infant mortality had been grossly exaggerated by the urban authorities in order to divert attention from the squalid conditions in town slums which were the real cause of most tuberculosis.[11] Town sanitary officials meanwhile mustered the resources of the new microbiology and health statistics to sustain the charge that it was the failure of rural producers to fully adopt the measures demanded by sanitary science which was chiefly responsible for flooding the towns with infected milk.[12] At stake in this conflict was the ability of the ascendant class of experts to establish a spatialized or territorialized process of ontological purification, in which the inflow of nonhuman substance from the country would be sanitized before its consumption in the towns. In this vision, the space of culture, of modernity and order, was to be insulated from the space of hybrid encounters, interspecies relations, contingent nature and disorder.

In this context, the discourse of 'clean milk' represented itself as an attempt to foster an attitude of public service among dairy farmers in the interests of urban consumers, but it was also an expression of the interests of the ascendant class of experts and their attempt to redefine the emergent late-modern public sphere in their own terms. Hence the definition given to the general or public interest by a specific class was in fact an assertion of its power and authority; it was an affirmation of the social importance of that class. This had an epistemic dimension, in that the discourse of a separate and disinterested science and the formation of the late modern public sphere around the authority of scientific experts were intimately connected. It was not just about the cleanliness of milk and the health and safety of consumers, it was about the ontological, epistemic and political organization of modernity itself. In order for the discourse of milk hygiene to function as the legitimation strategy for the assertion of expert power-knowledge, it had to appear to transcend sectional interests; thus clean milk was spoken of as a socially undivided collective 'good', as 'a subject which directly affects every class of a community'.[13] In this way, milk hygiene provided discursive terrain for the historical interpellation of the late modern 'social' body.

Contamination, dirt and disease provided the 'other' against which the sanitary authorities could define their own 'social' and legitimize the extension of the disciplinary mechanisms, political technologies and regimes of surveillance at their disposal; they were the mirror in which this public sphere glimpsed its own inverted reflection. References to the public interest by the sanitary authorities of the time in their reports, recommendations and articles, especially in the context of the milk supply, may be read as eulogies to their own burgeoning authority, to the widening scope of their powers of inspection and surveillance, and to the capacity of their technical knowledge and expertise to control a range of material social processes. This involved a conflict with farmers, sometimes latent, often explicit, in which a particular vision of 'civilization', 'culture' and the 'social', in contradistinction to 'nature', was hammered out for late modernity. Such conflict is clearest in the almost obsessive concern of the municipal authorities with the inspection of milk coming into the towns by rail from the country, in which the boundaries of the urban area came to signify much more than a purely geographical frontier.

Although there was continuing resistance from farmers to the idea that tuberculosis was zoonotic, the terms of the debate shifted dramatically in the years between the mid-1880s and 1909, and those who continued to deny the identity of human and bovine tuberculosis became increasingly isolated. As the question was gradually resolved in terms which served to reinforce the authority of expert technical knowledge, the response to zoonotic disorder could no longer be a concerted effort to deny or ignore the transgression it represented, as in the earlier period, but instead had to be founded upon a material renegotiation. Thus the ground was laid for the spread of systematic testing of the milk brought into the cities by rail, together with farm inspection and the tuberculin testing of all cattle at farms found to be producing tuberculous milk. This was a shift in the *modus operandi* of humanist discourse, as it became increasingly articulated within scientific and technical knowledge-practices, rather than within the traditional knowledge of working farmers.

Many smaller farmers were reluctant to co-operate with the sanitary measures being imposed upon them, not only because of their lingering scepticism as to the contagiousness of tuberculosis, and especially towards its zoonotic communicability, but also because of their lack of free capital with which to pay for further sanitary measures. Many were struggling in the face of a rapidly changing industry and could by no means afford to make the necessary investment.[14] As inspection regimes tightened and sanitary legislation multiplied at an accelerating rate, it was these farmers who frequently bore the brunt of the new disciplinary apparatus, as they found themselves progressively priced out of business by the spiralling demands of 'clean milk'. In this way sanitization did much to create the conditions in which the larger enterprises could dominate. This economic situation in turn fuelled the resistance of smaller farmers to the encroachments of official inspection and expert surveillance. In this struggle, the experts were eventually able to gain the advantage of consensus, allowing them to act in a far more concerted and effective manner than their critics, who were typically isolated and fragmented in comparison and often underrepresented in producers' associations.

Such consensus was not quick to emerge, however; it was preceded by over a decade of vociferous disagreement and rival claims, the beginning of the end of which can be traced to the deliberations of the 1896–98 Royal Commission on Tuberculosis. Previously this disease had been avoided or marginalized in official investigations into cattle diseases and milk hygiene. It was conspicuously absent from all the early legislation against cattle diseases, for example, while far less common infections such as scarlet fever were included. This was partly because the clinical symptoms of tuberculosis were notoriously hard to identify and difficult to distinguish from various other diseases; thus tuberculosis was something of a grey area until the bacillus was identified microscopically, and even for some time after this. But it was also because, as Keir Waddington has noted,[15] official responses to cattle diseases were organized around a separation of the 'economic' problems of 'agricultural' diseases from the 'public health' problems of unclean milk. Tuberculosis flouted this divide materially and was therefore able to thwart systematic consideration for some quarter of a century after other diseases began to be specifically combated in legislation; it quite literally slipped through the gap created by the separation of human health matters from animal health matters, a domain of social relations between subjects and a domain of objective relations between objects.

The inauguration in 1899 of the National Association for the Prevention of Consumption and Other Forms of Tuberculosis was therefore a significant turning point. It was based on the recommendations of the Royal Commission, which were as follows:

> Firstly, a systematic inspection of the cows in all dairies and cowsheds by the officers of the local sanitary authority within whose districts the premises were situated. Second, inspection, if requested, of any cows in any dairy or cowshed, wherever situated, by the authorised officers of local authorities within whose district that particular milk is supplied. Third, power for a medical officer of health to suspend the supply of milk from any suspected cow for a limited period pending a veterinary

> examination. Fourth, power to prohibit the sale of milk from any cow certified by a veterinary surgeon to be *actually* suffering from tuberculosis, or from such disease of the udder as in his professional judgement renders the animal quite unfit for the supply of wholesome milk. Fifth, the provision of a penalty for supplying milk for sale from any cow having obvious udder disease without the possession by the owner of a certificate that such udder disease is not tubercular. Sixth, the compulsory notification by all owners of cows of any disease in the udder (under a penalty). Seventh, proper separation and isolation of the diseased animals from the healthy ones, and lastly, eighth, the use of tuberculin as a detective agent by farmers and stock owners.
>
> (Hay Forbes, 1899, pp. 7–8)

The second of these recommendations was most radical in its implications. In the context of the now vast daily movement of milk by rail from rural producers to urban consumers, which had come to define the industry, it effectively meant empowering urban authorities to control the milk that was supplied to consumers within their jurisdiction. Thus what had begun as an opportunistic move by dairy farmers taking advantage of the shortage of milk in the cities in the wake of the 1865 cattle plague, had now been transformed in terms of the relations involved, so that the cities increasingly dominated the milk trade and sought to control rural production through the mechanism of sanitary regulation. Municipal health authorities now sought to build an apparatus of sanitary surveillance by constructing a cordon sanitaire encircling the city, a boundary maintained by the systematic collection and testing of samples of milk and by the inspection of cowsheds and milkshops. The boundaries corresponded to the limits of municipal jurisdiction, but also involved regular forays into the areas beyond in the name of farm inspections. These were more than public health measures; they were also ontological boundary-making practices, spatialized re-inscriptions of humanist dualism over the agnostic flows of hybridity and disorder etched by zoonotic tuberculosis. By controlling and progressively eliminating tuberculosis materially, the capacity of this nonhuman 'agent' to disrupt the ontological security of modernity was being made invisible.

Islands of 'culture': the sanitary city

> If the rise of farming was a bonanza for our microbes, the rise of cities was an even greater one, as still more densely packed human populations festered under even worse sanitation conditions. Not until the beginning of the twentieth century did Europe's urban populations become self-sustaining: before then, constant immigration of healthy peasants from the countryside was necessary to make up for the constant deaths of city-dwellers from crowd diseases.
>
> (Diamond, 1998, p. 205)

The vision of a bounded and sanitized milk-consuming community found one of its earliest and most influential expressions in Manchester, which was the location of the pioneering milk testing and farm inspection regime that came to be known as the 'island system' or 'Manchester method'.[16] To a significant extent this was the product of collaboration between the Medical Officer of Health for the city, a Dr Niven, and Sheridan Delépine, a professor of bacteriology at the University. Delépine was a particularly zealous clean milk campaigner, who undertook to make the resources of his laboratory available to the municipal authorities for the testing of milk samples.[17] In contrast with many sanitary officials at that time, who were still inclined to moderate their views in order to avoid both antagonizing farmers and alarming consumers, Delépine was uncompromising in his pronouncements on what was wrong with the milk supply and what had to be done about it. He is therefore interesting for my purposes because he gave unusually clear expression to the logic of the emerging sanitary regime; for this reason I will examine his views in some detail.

Delépine proposed a standardized, centralized system of milk testing backed up by legislative authority and linked to regular farm inspections, with punitive powers to seize milk, close milkshops and halt the flow of milk from suspected farms. His view was that it would ideally be desirable for the state to implement this regime in a uniform manner, but frustrated by what he regarded as the government's intransigence on the matter, he came up with an alternative strategy, arguing that 'so long as the state remains passive, it is the duty of enlightened principalities to protect themselves against every possible source of disease'.[18] But his ambition to see a more general enforcement of the system remained, and he conceived of his method as a strategy for achieving this one town at a time, thus: 'My system has sometimes been called the Island system, but this does not indicate exactly its nature. Its character is one of *centrifugal extension* by the gradual addition to the area under control of neighbouring areas not yet under control.'[19] The effectiveness of the method in Manchester contributed considerable momentum to the campaign against tuberculosis nationally, so that within thirty years the vision of a joined-up web of municipal milk inspection underwritten by the state was close to realization. In what then did the 'island system' consist?

Essentially the innovation of Delépine's method was to draw a sharp division between town and country in sanitary terms and to treat all rural producers as potential suppliers of tuberculous milk, by adopting a logic of siege or plague. This sanitary discourse inscribed a powerful spatial imaginary in which the city was counterposed to the country in terms of relative bio-onto-logical security, an imaginary relation which became an administrative reality. Previously, although borough authorities had powers to inspect farms and milkshops within their jurisdiction, they had no such powers with regards to farms beyond their sphere of influence.[20] The Manchester Milk Clauses of 1899, which were based upon the Royal Commission recommendations and which Delépine was instrumental in shaping, transformed this state of affairs dramatically.[21] From that time forward, samples of milk coming in from various farms in Lancashire, Cheshire, Derby, Stafford and Salop were collected at the railway stations, or elsewhere at the gateways to the city, by the food and drug inspectors. These samples were then sent to the bacteriological laboratory for examination. Any farm found to be sending milk which caused tuberculosis by inoculation was inspected by the city's veterinary surgeon, who examined all of the cows on the farm and took a sample of milk from each cow having a visibly diseased or 'suspicious' udder. These samples were then tested at the laboratory. Meanwhile the medical officer of health requested the farmer to isolate the cows yielding tuberculous milk, to cease sending the milk of these cows to town, and to have the cows with tuberculous udders slaughtered in the presence of the veterinary surgeon of the Corporation.[22]

Apart from the various processes of testing, both the testing of milk and the testing of cows made possible by the recent availability of tuberculin as a diagnostic agent, the method was one of disciplinary organization; it was a system of precise rules designed to engender a certain kind of order.[23] To illustrate this it is worth quoting Delépine's summary of his method at length:

1 Division of the whole country into a number of well-defined *administrative areas*, each provided with an efficient staff of veterinary inspectors.
2 *Marking and registration* of all cattle for purposes of inspection and notification. (This would, of course, mean the inclusion of tuberculosis among the recognised contagious diseases). *Systematic periodical inspections* of herds and cowsheds within each administrative area. Testing with tuberculin of all cattle.
3 *Isolation* of all tuberculous cattle. *Disinfection* of all sheds which have been occupied by tuberculous cattle.
4 *Immediate slaughter* of all cattle in an advanced state of tuberculosis (including all cases of udder tuberculosis).
5 *Fattening of all animals not in an advanced state of tuberculosis* for the meat market. All cattle should be slaughtered in *public abattoirs*, so as to ensure *thorough meat inspection*.
6 *Compensation* during one year or two for losses incurred by owners of cattle, *except in cases where there had been gross or culpable neglect*. After this *transitional period of compensation* the presence of a case of advanced tuberculosis in a cowshed ought to render the owner liable to a penalty.
7 *Testing of all cattle brought into the area*, so as to prevent the introduction of fresh sources of tuberculosis. No cattle above 6 years of age to be imported.
8 All the *milk from tuberculous cattle to be boiled* before use.
9 Control of all foreign dairy produce, so as to enforce the same standard of purity as in the case of home produce.

> 10 To avoid a sudden national depression in the dairy industry, I further suggest *that the enforcement of all these measures should not be simultaneously carried out all through the country*, but that they should be at first enforced in a certain number of well demarcated administrative areas, to which, year after year, other districts should be added, until the whole country was under administrative control.
>
> (Delépine, 1911, p. 31)

This was a system of exacting rules for the conduct of conduct, designed to eliminate the possibility of infection via the establishment of sanitary order, of discipline. It is strikingly reminiscent of Foucault's description of the method deployed in tackling the plague, with its extraordinary mix of analytical precision, calculation and corporeality.[24] And note that the deployment of order as a means of combating zoonotic disease suggests that these phenomena do not exist on different ontological levels, as modernity maintains, the one being a natural phenomenon and the other social. On the contrary, both are here revealed as forms of relation between material configurations, and these relations are not *a priori* incommensurable; it is humanist discourse and practice which render them so. In a sense, the Victorian sanitarians and hygienists were better able to see this than we are today, albeit implicitly, for they routinely conflated the social and the natural; they regularly spoke of disease as a form of social disorder, to be fought through social measures. This was not purely figurative. What they were historically unable to see, however, and what we are now just beginning to be able to glimpse, is the vast work of purification which engenders this anthropocentric 'social' in the first place, by extricating it from a nonhuman 'nature' that both encompasses and permeates it.[25]

Between 1896 and 1906 alone over 7,000 samples of milk were tested in the Manchester laboratory.[26] The impact of this work was significant and was quantifiable in the statistics which emerged from the testing process:

> The administration of the Manchester Milk Clauses has had a very marked effect upon the state of the farms supplying milk to Manchester . . . it will be noticed that, with one slight break in 1903, the improvement has been very steady, and that at the present time the number of farms producing tuberculous milk is less than one half of what it was in 1896 to 1900.
>
> (Delépine, 1908, p. 3)

> The effect of inspection has in many cases been remarkable, probably due to the rapid removal from farms liable to inspection of a large number of tuberculous animals. This is the only way in which I can account for the considerable diminution which has taken place in the proportion of tuberculous milk sent from several districts immediately after inspection had begun in these districts. This improvement was quite out of

> proportion with the number of tuberculous animals actually discovered by the inspector.
>
> (ibid.)

This amplification of effects is characteristic of the productivity of discipline, the way in which it functions to increase utility.[27] Indeed, it is useful to regard the ensemble of milk testing and farm inspection as an apparatus of panoptical surveillance of rural producers by urban sanitary officials, operating as a mechanism for the production of increased utility by obliging farmers to subject their own practices to discipline without the need for enforcement in a large number of cases; the constant possibility of punitive action was in itself enough to induce a widespread change in conduct. In this way fairly limited resources were able to precipitate behavioural change on a dramatic scale. It was not merely farmers who were surveilled, however, for I want to argue that this was also a means of ontological surveillance, a disciplinary mode of regulating the boundary between the human and the nonhuman. Thus the cows themselves became a surveilled population, as their potentially zoonotic agency, mediated through the flow of tubercular milk and borne along on the currents of commodity-exchange, was carefully contained and eliminated. As with all forms of rational disciplinary organization, however, there were unintended consequences, resulting from the magnification of effects upon 'externalities'. Chief among these was the impact upon localities which did not strictly regulate their milk supply in this way, for cattle that fell under suspicion tended to be quickly transferred to farms supplying their milk to areas which did not operate systems of milk testing and farm inspection. There were several high-profile cases in which farmers found to be supplying tuberculous milk to Manchester, upon discovery, simply began sending it elsewhere, or even using it to make cheese.[28] In this way, Manchester – like the humanist 'culture' of which it was a spatial and administrative expression – became a sanitized 'island' protected by a technico-political boundary.

The 'island system' was not the only method for the eradication of milk-borne tuberculosis in circulation at the turn of the century, but it was distinctive in being based upon a vision of total sanitary control of the milk supply by the municipal authorities, an ambition which emerges very clearly in the historical documents. Just prior to the adoption of his system in the Manchester Milk Clauses, Delépine set out the advantages of his method over various alternative approaches, and in particular over the method of combating tuberculosis by the partial removal of tuberculous cattle, which was considered by many to represent a more feasible compromise. He rejected this on the grounds that it did not provide 'for an efficient supervision by competent men of every beast in every district'.[29] This phrase is significant, for it makes explicit what was really occurring in terms of human–animal relations. Animals in the context of zoonotic infection were becoming regarded by the medical-scientific and political authorities as a threat to the modern order itself, to the coherence of humanist 'culture'; for this reason they were increasingly to be supervised, tested and recorded. Only by knowing animal bodies precisely and controlling them extensively could their potential for ontological pollution and disordering be managed. The other methods which might be employed against tuberculosis were inadequate precisely because they failed to secure such control. They could only ensure the elimination of the worst excesses of zoonotic infection and were not methods for achieving the total elimination of the disease.[30] Delépine similarly rejected any method relying upon voluntary notification by farmers, which many of his contemporaries regarded as preferable given the friction between veterinary officers and farmers which would result from any attempt to introduce official farm inspections.[31] But for Delépine, nothing short of absolute control would suffice in the face of zoonotic disorder; purity was by definition an absolute state. So it was either order and purity or disorder and contagion; there could be no compromise, mangle, or hybrid of the two:

> The work should be carried out *uniformly, simultaneously, and thoroughly, over well-defined, fairly extensive areas*, by a sufficient and efficient staff of inspectors, capable of inspecting *all the farms, of controlling the state of all the herds, and the importation of any cattle from outside.*
>
> (Delépine, 1911, p. 29)

> As soon as the work carried out in the first district had produced very marked and durable effects, one or more of the adjacent districts would be added to the controlled area, and *this would go on until the whole country was under control.*
>
> (ibid., my italics)

The totalizing language in these passages illustrates just how ambitious the Manchester authorities were in their efforts to secure control of the milk trade on the grounds of eliminating milk-borne tuberculosis. The word 'control' was frequently used by municipal officials in this connection, and it suggests more than mere intervention, signalling instead a framework of regulation systematic enough to take account of all of the contingencies in play. It suggests, in other words, the desire for a complete art or science of governance of the object of control, of the nonhuman animal and the various mediations of its corporeality. This was a new way of thinking; in practice, it meant an expanding system of urban sanitary discipline. A distinct kind of urban space was produced by this discipline; it was a bounded analytical space, a space of administered purity, increasingly defined and organized in terms of its being surrounded by an 'other', a 'nature', in which classifications were thwarted, identities mixed and boundaries crossed.[32] This is the space of 'civilization' and the frontier; it is the space of modern 'progress', of the 'cultural' order facing uncertainty and disorder on its borders.

A new sanitary paradigm: othering the nonhuman

> The new, continental, bacteriology was not about better drains or nutrition; it focussed on specific diseases, technical fixes and regulations against contagion. It was a very analytical, bureaucratic, *Prussian* science, but one which still echoes as part of our repertoire when thinking about public health.
>
> (Pickstone, 2000, p. 52)

The apparatus of milk testing and inspection as it developed in Manchester and other large cities was at once a technoscientific assemblage and a disciplinary mechanism for the creation of ordered, sanitary spaces, of purified 'cultural' zones.[33] It was part of a wider restructuring of humanist discourse during this period, in which new sanitary technologies and forms of discipline were increasingly deployed in the meticulous separation of the human from the nonhuman. Thus a shift in the *modus operandi* of humanism was expressed as a transformation in sanitary discourse and practice, from an earlier and self-consciously 'social' Victorian approach, which had taken the separation of urban 'public health' sanitation and agricultural disease prevention (underpinned by the human/nonhuman divide) for granted in its operating rationale, and towards an approach which perforce acknowledged the human and the nonhuman as materially continuous and laboured to orchestrate their separation by technical means. Hence, until the mid-1890s there was in almost all public discussion of tuberculosis a deeply ingrained and fundamental distinction between human-to-human and bovine-to-bovine routes of infection. The possibility of bovine-to-human infection was effectively made invisible or unthinkable by sanitary categories structured by a binary division between the first two routes; its specifically transgressive zoonotic logic was always evaded. In this way the separation of the human 'social' domain from the nonhuman 'natural' domain was mirrored, affirmed and reproduced in sanitary discourse, organization and practice.

It is tempting to say that this was dramatically shattered by the identification of the tubercle bacillus by the German bacteriologist Robert Koch in his laboratory in 1882, and by his initial conclusion – which he later rescinded – from a series of experimental inoculations, that the disease in cattle was the same as that in humans and therefore communicable.[34] Certainly this was a critical enabling factor in the demise of Victorian 'public health' in its existing form, but its impact was not felt immediately. Indeed, it took fifteen years of continuous dispute between microbiologists, pathologists, hygienists, clean milk campaigners, farmers' associations, milk sellers, dairy companies, consumers and sanitary officials before the implications of Koch's early laboratory work began to become compelling. It was never compelling in and of itself, but was only made so by being mobilized as part of an emerging politico-epistemic network.

By 1897, Delépine was able to argue that the fact 'that tuberculous infection of man by man is more common than infection of man by cattle, is no reason why one should postpone dealing with bovine tuberculosis 'till all human sources of infection have been dealt with.'[35] This was among the earliest criticisms of the dominant approach to tuberculosis under the existing 'public health' paradigm, which focussed upon 'social' measures for eradicating infection and was therefore unable to acknowledge that the dynamics of bovine-to-bovine infection and those of human-to-human infection were materially interrelated through the circulation of milk:

> Although the sanitarian and the agriculturalist have a common interest against tuberculosis, the special methods which afford a measure of protection to man are quite insufficient to protect cattle against the spread of the disease. The reverse is not true, for if bovine tuberculosis were stamped out by general measures the sanitarian would have nothing to fear from tuberculosis of cattle. The eradication of bovine tuberculosis is, therefore, a subject in which the hygienist is keenly interested.
>
> (Delépine, 1911, p. 9)

This reasoning lay behind the gradual erosion of the distinction between 'public health' sanitation and 'agricultural' disease prevention taking place at that time. This was both a response to the new human–nonhuman relations inscribed by zoonotic tuberculosis and an attempt to manage those relations, to govern them both materially and ontologically. The separation of human from nonhuman was not to be abandoned in this new paradigm, however, thereby ushering in some post-humanist era, for the shift was essentially an operational or methodological one, a change of means rather than a fundamental ontological transformation. Thus, whereas previously the human/nonhuman dichotomy had been enacted as a given within the distinction between public health and agriculture, this was now increasingly problematized materially, and had therefore to be re-articulated in the face of the new hybrid agent made visible by the microscope. Certain entities such as tubercle bacilli had become 'obligatory passage-points'[36] established by the emerging networks of expert technical knowledge, and which only technical knowledge-practices could now negotiate. Hence the work of governing these nonhumans was increasingly undertaken within these practices, and involved a certain kind of practical acknowledgement of the human–nonhuman continuity that it laboured to make invisible. In this sense these practices were deeply schizophrenic, for they were involved in the perpetual making-absent of their own conditions of necessity. The proliferation of these technical purifying practices was concomitant with the decline of the Victorian sanitary paradigm organized around a separate urban 'social', and the gradual emergence of a strategy at once more complex, double-sided, and circular. The age of technoscientific modernity or modernity proper had begun.

An example which nicely illustrates the nature of this shift is the changing practice of milk testing in Manchester. Until the 1890s, the favoured method in use for protecting consumers from infected milk was by 'the production of increased resistance to infection'. This meant improvements in housing, nutrition, clean water, air quality, hygiene and drainage, which it was believed would increase people's resistance to tuberculosis, for it was known that the weak, the undernourished, and those living in dirty and overcrowded conditions, were more likely to contract the disease.[37] This approach was typical of Victorian 'social' sanitation, which sought to control 'natural' forces through the manipulation of 'social' conditions.[38] Interestingly 'the social' in this strategy was a rather inclusive category, encompassing an array of material entities and human–nonhuman assemblages, and even aspects of the 'natural' environment, in fact everything perceived to be a 'vital' condition for human beings to function 'normally' as 'vital' subjects of modern urban society.[39] It was widely supported by farmers as the best approach to tackling tuberculosis, no doubt because it did not legitimize any form of surveillance of farms. As the anxiety associated with the question of milk-borne tuberculosis grew, the sufficiency of 'social' sanitation began to be questioned. It was criticized for its failure to tackle the disease 'at source', hence for its separation of 'public health' from 'agricultural' measures. The source of the problem was increasingly located as exogenous, within the nonhuman and the 'natural', and especially in the unregulated passage from the nonhuman to the human, rather than endogenously within the human social domain. Hence, while the 'social' approach took the autonomy of 'the social' for granted, this was precisely what was under threat from zoonotic disorder and which had to be actively entrenched by boundary-making practices. The various methods of testing milk for tuberculosis can be understood as responses to this problem; there were three types of tests, each with very different implications:

First, there were microscopic tests. These involved examining tiny samples of milk under a microscope for tubercular bacilli. Due to the central role of the microscope in the initial identification of tuberculosis, this was often assumed to be the most 'scientific' and reliable test. In fact, it was the least reliable, as the bacilli could be extremely difficult to detect in a sample of milk if they were as yet small in number.[40] This was compounded by the fact that the great majority of the milk being transported into cities was mixed milk, that is, it was the milk of many cows mixed together; hence the chances of detecting the bacilli in any given sample were minute.

The second type of test was the cultivation test. This involved encouraging the growth of bacteria from samples of milk placed on agar jelly. The rate of multiplication of the bacteria allowed the original numbers of bacteria in the sample to be calculated, which provided a measure of the relative cleanliness of the milk. These cultivation tests, the numerous tabulated results of which were a staple feature of the reports of the Medical Officers of Health for Manchester and Salford, were actually more relevant to the campaign for 'clean milk' than to measures to secure milk free from tuberculosis.[41] Along with tests which made use of a centrifuge to separate out and measure the amount of 'dirt' in the milk – which formed a sediment or 'slime' consisting of cells from the cow's udder; hairs from the cow's skin; matter from food, dung or litter clinging to the skin of the cow; worms, insects and arachnids; yeasts and moulds – these bacterial tests were really a test of the standards of hygiene obtaining during the processes of transportation and storage of the milk, hence a test of the conduct of the humans involved in conveying the milk, rather than of the diseased or healthy status of the cow.[42] This method of testing was therefore of a piece with a slightly different set of concerns, with a slightly older history, as to the 'purity' or 'cleanliness' of milk, variously defined. They were part of a 'public health' paradigm ill equipped to meet the challenge posed by the emergence of milk as a medium of zoonotic contagion.

The third kind of tests were inoculation tests, which involved injecting a small quantity of milk into a nonhuman animal of some kind, usually a guinea pig, and then simply waiting to see whether the animal developed symptoms of tuberculosis.[43] This was the most reliable method, but it could take up to three weeks to obtain a clear result, which militated against its preventive use. Hence even this kind of test was not a panacea for the problem of how to obtain a milk supply free from tuberculosis.[44] Indeed, no form of laboratory testing could in and of itself achieve the desired result.[45] When combined, however, with an additional system of surveillance, in the form of regular farm inspections and tuberculin testing of suspected herds, inoculation testing proved highly effective, for it was part of an approach which for the first time bridged the gap between municipal public health and agricultural disease prevention.[46] In this way, the emergence of zoonotic tuberculosis led to the development of a form of sanitary organization which was no longer 'social' in the earlier sense, but rather 'purifying'; it no longer operated on the prior assumption of a separate human 'social', but upon a heterogeneous human–nonhuman collective, from which it laboured to *engineer* a separate human 'social' and to make this 'social' appear as autonomous and exclusively human as ever.

The emergence of this new sanitary paradigm saw a fragmentation of the older distinction between 'clean' and 'unclean' milk, as it emerged in the light of new methods of testing that there were multiple ways in which milk could be 'unclean'. The tests were undertaken by an ascendant class of scientific professionals employed by municipal authorities, consisting of food and drug inspectors, medical officers of health, pathologists and bacteriologists. Each kind of test was more than a technical practice, for it inscribed a specific vision of agency, of ontology, of subject and object, social and natural, with commensurate attributions of responsibility, of justice and morality. In this way the laboratories where the tests were conducted became key sites of ontological labour in which modernity was perpetually engineered. Butterfat tests, for example, in which the percentage of fat in milk was measured, were established against considerable opposition as tests of the 'authenticity' of milk, of whether it had been adulterated, with possibly dire consequences for farmer or retailer if a certain standard of composition was not reached. The effacement of the 'agency' of the cow's body, the so-called 'appeal to the cow', with regards to such tests was achieved only after intense struggle between dairy farmers and the prosecuting authorities. The cultivation test, similarly, in which the bacterial content of milk was determined, was used to establish human culpability on the assumption of poor hygiene during milking or transportation.[47] Above all, though, the inoculation test for tuberculosis was used to identify farms from which tubercular milk had been sent into the municipal area. This enabled individual farmers to be identified and disciplinary action taken where it was believed that a farmer had knowingly sold tuberculous milk.[48] Thus the testing regimes were not merely means to ensure the 'purity' of the product or the 'safety' of the public, but by inscribing models of agency, objectivity and responsibility, they were also disciplinary techniques for the policing of the unstable boundary between the human and the nonhuman, between 'culture' and 'nature'; they were boundary-making and purifying practices.

While other towns and cities were following Manchester's example by implementing similar testing and inspection regimes, there was continuing scepticism within agriculture itself as to the relationship between human and bovine tuberculosis. There was an epistemic gulf between the expert view of pathologists and veterinary inspectors, facilitated by the new form of visibility engendered by the microscope, and the practical view of dairy farmers and their producer associations, who tended to play down the threat from tuberculosis, and many of whom continued to deny its transmissibility to humans well into the twentieth century. This schism both paralleled and fuelled the wider political and ontological struggle between urban health authorities and rural milk producers. Such antagonism was very much in evidence at the Dairy Conference of 1909, which brought together sanitary experts and working dairy farmers and was representative of the tensions pervading the milk industry at that time, as it grappled with zoonotic tuberculosis. The lines of division were not always drawn, as one might expect, according to a hierarchy of expertise, for a significant minority of experts remained unconvinced of the significance of tuberculosis transmitted through milk, and a minority of farmers were highly exercised by the problem. To give an impressionistic sense of the conflicts which pervaded both the conference and the industry more widely, it is worth examining the views of some of the delegates in more detail.

In his speech to the conference, Professor J. P. Sheldon accused the urban health officials of libelling milk producers by repeating

> the persistent charge, brought by officials here and there, by certain journalists, by men who ought to know better, and, indeed, who do know better, that milk from the shires is responsible for much, if not all, of the infant mortality from tuberculosis that occurs in urban districts, chiefly in those that are most thickly crowded with human beings. It is obvious

> that such a charge is calculated to do enormous mischief in more than one direction . . . These officials are silent about the numberless infant lives which milk from the country has saved from destruction.
>
> (Sheldon, 1909, p. 28)

Others were prepared to go further; J. T. Horner was 'of the opinion that the danger of consumption being conveyed by milk from the bovine to the human being has been greatly exaggerated'.[49] While F. J. Lloyd, the Consulting Chemist to the British Dairy Farmers' Association, received a burst of spontaneous applause for his declaration that he was 'one of the few men in England who believed that tuberculosis in cattle had nothing to do with the disease of the same name in human beings'.[50] Archibald MacNeilage, a delegate from Glasgow, which had also been quick to adopt farm inspection, poured scorn upon this conviction, charging that if 'Lloyd and the gentlemen who agreed with him really believed that, let them, in the interests of the public, drench themselves with milk containing the germs of bovine tuberculosis, and thus demonstrate the sincerity of their belief'.[51] Against this, A. J. Mullins, representing the Somerset, Wiltshire and Dorset Farmers' Association, adhered to the farmers' traditional view that tuberculosis was not even contagious between cows; he insisted that there were but three causes for tuberculosis: 'First, it is hereditary; secondly, it is due to the atmosphere in which the cows were kept; thirdly, it is brought on by drafts causing a chill.' In something like a theory of 'intelligent design', Vernon B. Chalk posited the view that 'the cow's body contained the finest system of filtration organs that could be constructed'. He therefore demanded to know whether it was 'possible for nature to construct a cow to feed her people and at the same time to use the milk for the purpose of destroying them?'[52] This was rubbished by the Consulting Veterinary Surgeon to the BDFA, who rather than using the metaphor of an intelligent nature deployed the language of industrial pollution, stating that cows with symptoms of tuberculosis were 'manufactories of tubercle germs, which they distribute all round them, contaminating drinking water and fodder, and even the air itself'.[53] The fierceness of the controversy and the complete lack of common ground underline the significance of the epistemic shift that was taking place.

Even as late as 1909 many dairy farmers were still denying the communicability to humans of bovine tuberculosis. This denial was also resistance. For a long time the disease had been known to farmers under various local names, such as 'the grape disease', 'pearl disease', 'wasters', 'piners', or 'snorters', according to whether the lungs, the bowels, or the throats were the primary organs attacked. It was only with the development of bacteriological science that these disparate symptoms were recognized as belonging to a single disease. Previously there was no disease known as 'tuberculosis', but only a host of folk ailments believed to be precipitated by a variety of ill-defined conditions, by heredity, or by the whim of fate. Critically, the establishment of a singular 'tuberculosis' was not initially an 'empirical' development in the strict sense, but an analytical one; it depended upon a diagnostic abstraction, the postulation of a sole causal agent, detectable only in the laboratory. This new way of knowing dramatically undermined the traditional knowledge of dairy farmers, which was grounded not in such technoscientifically enabled abstraction but in practical experience and empirical observation. Farmers suddenly found themselves facing an unknown enemy, about which they were obliged to learn from experts outside of agriculture. It is of secondary significance that microbiological knowledge would in time enable farmers to better combat the disease; to tell this kind of story is merely to rehearse the modernist discourse of historical 'progress'. Of more importance for a symmetrical account is the rupturing of the existing networks of knowledge-practice by the unity of this new epistemic object, this creation of technoscientific modernity. Although their ability to combat actual tuberculosis had hitherto been severely limited, farmers had been able to manage the ontological insecurity attendant upon the disease through their local classifications, categorizations and beliefs; centuries of farming practice had conferred a degree of control over the unseen and the nonhuman. This sense of ontological security was ruptured by the tubercle bacillus which, combined with microscopes, testing procedures and laboratories, effectively concentrated the power of classification, the power to know and control the nonhuman and to regulate its boundaries with 'society', in the hands of technical experts.

Knowledge and disease: the ontological politics of aetiology

> Against the plague, which is a mixture, discipline brings into play its power, which is one of analysis.
>
> (Foucault, 1991, p. 197)

Just as 'piners', 'wasters' and 'pearl disease' were the creations of farmers, so 'tuberculosis' can be understood as a creation of the experts who detected and studied the bacillus, disseminated their findings, and helped frame the regimes of inspection which, while predicated upon the materiality of this object, were simultaneously the 'theatres of proof' that endowed it with social and political force. They were involved not merely in the communication of newly acquired knowledge of a transcendentally real object, but in the production of that object as 'real'. This in turn provided the rationale for the authority of the local officials charged with eliminating the disease by reshaping practice in line with the new 'reality'; existing practices were no longer enough to manage the disorder, either ontologically or corporeally. In this sense tuberculosis was deployed against rural farmers as a politico-ontological device by the sanitary authorities; it was a political weapon. In this light, the resistance of farmers to the claim that tuberculosis was communicable from cattle to humans, which was maintained by many for some time after the evidence had become compelling, can be understood as more than just 'rural backwardness' or conservatism, but as an attempt to resist the discursive positioning of urban experts as the guardians of 'culture' relative to farmers as unscrupulous outsiders located on the margins of modernity.[54] Having said this, the resistance of farmers was also enabled and fuelled by the persistence of disagreement among the experts themselves. A key turning point in this was the Koch controversy.

The publication of Koch's identification of the bacillus and the demonstration of its presence in tuberculous lesions and products obtained from humans, cattle, swine and other animals, together with the experimental demonstration of its pathogenic action, had appeared to represent an unanswerable confirmation of the theory of the unity of tuberculosis, the postulation that one disease was responsible for diverse observable symptoms in different parts of the body.[55] In 1884, Koch had further stated that bovine tuberculosis was identical with human tuberculosis, and could therefore be transmitted to humans by the consumption of the flesh and milk of tuberculous animals. Though there were still dissenting voices, the authority of Koch's international reputation helped to ensure that, on the whole, pathologists and bacteriologists became increasingly united in the belief that tuberculosis was communicable from nonhuman to human. Such unity was soon shattered, however, by Koch's sudden reversal of his position.[56] At the British Congress on Tuberculosis in July 1901, Koch caused astonishment by declaring that:

> I feel justified in maintaining that human tuberculosis differs from bovine, and cannot be transmitted to cattle. I should estimate that the extent of infection by the milk and flesh of tubercular cattle, and the butter made from their milk, as hardly greater than that of hereditary transmission, and I therefore do not deem it advisable to take any measures against it.
>
> (Cited in Delépine, 1911, p. 4)

It is something of a historical curiosity that Koch should have abandoned his earlier views so abruptly, and in such a way that the debate in Britain on the communicability of tuberculosis, which was by then almost settled, was thrown into disarray, giving the sceptics new impetus and prolonging the debate by another ten years. The Royal Commission on Tuberculosis set up in 1904 began to conduct exhaustive enquiries in an attempt to either confirm or refute Koch's conclusions. Meanwhile those farmers and dairymen who had always maintained their denial as to the zoonotic communicability of TB were boosted in their struggle against the local authority experts, who now appeared equivocal and divided.

The reason for Koch's abrupt reversal appears to have been his observation that there were morphological differences between the bacilli in humans and in cattle, leading him to the conclusion that these were different diseases. Even so, as a renowned bacteriologist he would almost certainly have been aware that such morphological variations did not in themselves necessarily indicate distinct species. In any case, this high-profile reversal temporarily rehabilitated the idea that the boundary between the human and the nonhuman was in fact more robust and less permeable than the majority of microbiologists had been suggesting. Thus division among the experts gave a brief reprieve to the fatalistic form of humanism that had preceded the identification of the tubercle bacillus and which was still the worldview of the majority of farmers. This was short-lived. The Royal Commission on Tuberculosis, set up in the wake of Koch's reversal to either confirm or deny his conclusions, eventually reasserted the unity of tuberculosis and its communicability from cattle to humans.[57] After ten years of investigations, its report was unequivocal in its conclusion that 'Mammals and man can be reciprocally infected with the disease.'[58] It further affirmed that

> In a considerable proportion of the cases of the human disease the lesions contain, and are caused by, bacilli which are the cause of tuberculosis in cattle. In all such cases the (human) disease, therefore, is the same disease as bovine tuberculosis.[59]

Despite the apparent clarity of this conclusion, the picture was becoming more complex. The various separate folk conditions having been identified as a singular disease, and the identity and communicability of human and bovine tuberculosis asserted, further investigation had then muddied this early clarity by revealing that in fact there *were* differences between human and bovine tubercle bacilli, differences in terms of both morphology and effects, relative virulence and pathogenicity. Yet these differences did not signify the incommunicability of the diseases, as Koch had believed, for bacilli of both types were present in both human and animal cases of the disease. The identity of human and bovine tuberculosis then was neither unambiguously confirmed nor denied, certainly not in the kind of black and white terms that sanitary reformers had wanted. Instead the disease thwarted the simple identity/difference binary by proving to be an awkward and fuzzy hybrid of the two. The bacilli were distinct in certain significant respects, but they were nevertheless to be regarded as a single species. It had been taken for granted that the unity and the communicability of human and bovine tuberculosis were synonymous, but now these two variables became partially independent.

The common distinction between human and 'animal' tuberculosis was also unravelling, for it was found that the comparative effects of the human and the bovine bacilli varied widely in their effects upon different non-humans. This made a methodological nonsense of the experimental use of a single species, typically the guinea pig, to affirm either their identity or non-identity. Thus the long-standing question in terms of which the debate had been widely understood, namely 'whether tuberculosis in animals and man is one and the same' was now rendered unanswerable, owing to the meaningless generality of the categories in terms of which it was framed. Tuberculosis itself thereby served to break down the humanist binary of human/animal, with its unacknowledged assumption that there was such a thing as 'animals' in the world, or in other words that there was some generic nonhuman 'animal' which might be meaningfully counterposed to humans. Crucially these epistemic destabilizations were not contained within an abstract realm called 'science', but were replete with implications for the wider organization of modernity. They demanded a reorganization of the prevailing ontological picture of the world, the relationships between its subjects and objects, and the place of humanity within it. This was accomplished within a political reorganization of the disciplinary systems of containment, separation and purification through which the relations between the subjects, objects, agents and entities classified in this ontology were inscribed, at once materially and discursively.

At the turn of the century then, the notion of a singular cause for tuberculosis was still contestable; it had been established under laboratory conditions, but had not yet permeated farming practice or acquired the monolithic status of a 'fact'. At first glance the identification of tuberculosis may seem to represent a vindication of 'the scientific method', which is to say the belief in reduction, reduction to single definitive causes and definitive or 'diagnostic' symptoms. This ontology of exclusive causes with its rejection of plurality is deeply monotheistic in its form, revealing the religious 'spirit' of modern scientific rationalism. But the search for sole causes was frustrated at every turn by proliferating complexity, and was only part of a wider reconfiguration of the scope and limits of agency and of what kinds of entities were endowed with it. Thus the identification of a sole direct cause of tuberculosis was paradoxically accompanied by a renewed emphasis upon numerous supposed 'accessory causes', as follows:

> Now, among accessory or contributory causes of tuberculosis are, first of all, heredity, in which consists the perpetuation of certain typical constitutions favourable to the growth of this particular bacillus when introduced; then we find climate, certain geological conditions, the influence of certain heating foods, the keeping of animals in small, cramped, dark, badly ventilated, and badly drained sheds, and lastly we find tuberculosis spread by contagion, chiefly from the particles of dust contained in the inhaled air and from infected food.
>
> (Hay Forbes, 1899, p. 5)

The priority given here to heredity is clearly a concession to the pre-microscopic ontology of farmers, who preferred to believe that heredity rather than contagious transmission was at the root of their problems, and that they could therefore breed their way out of cattle disease without having to invest heavily in new equipment or submit to testing and inspection. The language of accessory causes therefore marked an attempt to reposition the farmer as an 'agent' able to 'act' to prevent tuberculosis, thus recuperating some of the capacity for action lost to the microscope, to experts and to the bacilli. Though the bacilli had indeed partly usurped the farmer's claim to 'agency', this quickly became incorporated into a redefinition of the 'objective' context and limits of the agency of farmers, undertaken by the ascendant class of experts. This restoration of a transformed and bounded 'agency' for farmers was necessary to consolidate expert authority, which ultimately relied upon farmers' capacity for action.

There is then a remodelling of 'agency' inscribed within the early microbiological accounts of tuberculosis, in which purposive action upon nature is progressively dissociated from practical wisdom and traditional methods, until the farmer has little choice but to be guided by the dictates of expert knowledge. The human monopoly of 'agency' presupposed by the notion of 'culture' had been ruptured and destabilized by the interventions of the tubercle bacilli and the microscope; it could only be recovered by the redrawing of 'agency' such as to redefine the boundary between the two domains. The impact of nonhumans, hybrids, and zoonotic entities meant that this remapping could only be accomplished within assemblages of technical knowledge-practices capable of materially renegotiating the disorderings of which they were an intrinsic part. Hence the processes of purification central to the organization of modernity became increasingly the preserve of those with access to specialist knowledges and technologies; an epochal asymmetry was created which was above all ontological.

Experts then were being relied upon for more than just the practical utility of their knowledge, for intrinsic to this usefulness was their growing power to give order and definition to an increasingly disordered world, to know and categorize unseen forces. This epistemic shift was political in a myriad sense; it marked the emergence of a new paradigm of governmentality, a series of discursive shifts through which new ontological boundaries and identities were inscribed upon phenomena, hence a new anatomy of power rendering possible new techniques of control. But it was also more concretely a transformation in the relationships of power, knowledge and interests implicated in these techniques and discourses, crucially a decisive shift in the power relations between practitioners such as farmers and the network of technical experts and officials charged with the work of policing the boundaries of humanist ontology for modernity. Armed with the microscope and the laboratory, with experimental methods, with statistics and surveillance, these experts were involved in an incipient ontological politics, which positioned them increasingly at the heart of the government not directly of human bodies and urban life, which is to say of 'the social', but over nonhumans, over the cosmology of 'nature' itself. To govern modern 'society' it is necessary to govern its opposite, that is, to define, sanitize, and control its boundaries with 'nature'; hence the ascendancy of scientific expert knowledge and authority was integral to the realization of humanist modernity.

Conclusion: orchestrating 'the social'

The period from around 1890 to 1920 witnessed some important changes in the sanitary regulation of the British dairy industry, which laid the ground for the economic transformations of the following decades. At the centre of these changes was a paradigm shift in urban sanitary organization; this was precipitated by the problem of zoonotic tuberculosis conveyed into the cities by milk, but it had wider significance in terms of the mode in which humanist ontology was actualized within the political technologies of sanitary discipline. The earlier Victorian period had seen an ontologically 'social' (albeit discursively medicalized) approach to urban sanitation, in which the ontological separation of the human–social from the nonhuman–natural was expressed in the organizational dislocation of 'public health' sanitation from 'agricultural' disease prevention. This separation was gradually undermined by the growing evidence that tuberculosis was the same disease in humans as in cattle, and transmissible from one to the other through the medium of infected milk. Thus the zoonotic 'agency' which emerged from a network of heterogeneous relations between the cow, the microscope and the bacilli, effectively disrupted the human/nonhuman dualism of the existing sanitary paradigm; it demanded a restructuring before the ontological disorder could be managed.

A new 'purifying' sanitary paradigm emerged in response, and this was 'social' in a different way. In the old system, the 'social' had been enacted in a theological mode as a given, that is, as though it simply existed in the world and therefore had to be reflected in the organization of sanitary practices. In the new 'purifying' paradigm, the 'social' was something which had to be materially and technically orchestrated in the face of increasingly visible hybridity, zoonotic transgression, heterogeneous mixture and social-natural interconstitution. This orchestration had become a historical condition for the enactment of 'the social' and its reproduction as a given domain; it was now the ontological-political infrastructure enabling 'the social' to continue to be experienced and lived as a reality *sui generis*. The transformation therefore amounted to a restructuring of the mode of operationalization of humanist discourse, which is to say, of the humanist technologies through which the 'social' is produced by perpetually othering the nonhuman.

This shift was intricately connected to the ascendancy of experts and of expert knowledge, which was progressively established during this period. The identification of the tubercle bacillus as the singular cause of tuberculosis, hence of tuberculosis as a singular disease, was not the sole 'cause' of this ascendancy, but it was the critical element in a network of relations which increasingly enabled expert knowledge-practices to dominate over the traditional and experiential knowledge of practitioners such as farmers. Other key nonhuman elements in the emerging network were the microscope, the bodies of the infected cows and their susceptibility to the disease, the tissues and organs of laboratory animals such as guinea pigs, and the organic properties of the milk itself. This network was necessarily contingent, messy and uncertain; hence it was more than a decade after the initial identification of tuberculosis before it was able to coalesce into an effective historical conjuncture. The intervening years were characterized by widespread epistemic conflict, expert disagreement, paradigm instability and resistance from farmers. Once it became hegemonic, however, this network precipitated an epistemic shift which was also a shift in the ontological politics of modernity itself.

In this shift, the work of engendering 'the social' through a meticulous separation of the human from the nonhuman was progressively concentrated within technoscientific knowledge-practices, which increasingly became the locus of the ontological ordering processes constitutive of modernity. Hence those with access to specialist knowledges and technologies were empowered to give definition to the modern 'social' by knowing and classifying an unseen 'nature'. In the same process, traditional and practical knowledges were conversely stripped of their capacity to give ontological definition to the world. In terms of human–animal relations this shift in ontological politics meant that humanist discourse was increasingly insulated from those who actually interacted with nonhuman animals as co-producers on a daily basis. Hence the persistent problem of animals, which is to say the problem of their hybridity and of the tendency of human–animal encounters to undermine humanist categories and ordering processes, was now to be managed by means of the progressive distancing of these inter-corporeal encounters from 'the social'.

There was a spatial dimension to this distancing, as the city was increasingly imagined in sanitary discourse as a besieged island of bio-onto-logical security, as an enclave of 'culture' surrounded by the insecurities of a zoonotic 'nature'. This was materially enacted with the construction of a regime of milk testing and farm inspection, which effectively formed a sanitary cordon around the city, and which involved extensive surveillance of farming practices by urban sanitary authorities. Thus these problematic practices on the boundary between the human and the nonhuman were both distanced from and disciplined by 'the social city'. This discipline at a distance was also a means of surveillance of the cows themselves, a political technology by means of which their disruptive 'agency', their capacity to make a difference, and specifically to make such differences as were erosive of humanist ontology, was contained, managed, and as far as possible made invisible through the imposition of sanitary discipline as a form of ontological ordering. Humanist separation and purity were thereby re-established against the mixture and impurity of the disease. The nonhuman 'agency' in question was an emergent property of a network encompassing the laboratory and the microscope as well as the cow. Thus the imposition of discipline meant constructing a different network from these elements which re-articulated them within the humanist ontological architecture. As technical knowledge-practices were deeply involved in the emergence of this network, having effectively created zoonotic tuberculosis as an epistemic object and rendered it an obligatory passage point for subsequent ontological discourse, it was overwhelmingly within these practices that this work of purification had thenceforth to be undertaken.

There was a circularity to this, for these expert knowledge-practices necessarily involved a certain kind of acknowledgement of the continuity of the human and the nonhuman; this was integral to the work of separation itself – the recognition that human and nonhuman were not separate but constitutively interwoven. In this sense these practices were systematically self-negating; they were continually involved in the making invisible of their own conditions of necessity. This contradiction was possible because the conception of the experts in question concerning their own practice was shaped by the humanist separation of 'the social' from the 'natural' which these practices themselves engineered, and was therefore consistently non-reflexive in this respect; hence they were able to carry out the humanist work of purification without simultaneously undermining it. Thus the inoculation testing of milk, the tuberculin testing of cows, the inspection of farms and the isolation and slaughter of diseased animals, were not just sanitary practices or public health measures, but were also ontological-political technologies, through which the human–animal continuity inscribed by zoonotic tuberculosis was made invisible as a condition of reproduction of 'social' modernity.

5 Milk and modernity Part II

Measurement, rationalization and control

> The success of our agriculture depends largely on the cheapening of production, and to cheapen production we must make sure that every unit is worked to its fullest capacity.
>
> (Lawson, 1913, p. 58)

> These daily records of milk production tell the farmer the exact truth about his herd.
>
> (Newsham, 1913, p. 17)

Introduction: purification and rationalization

In this chapter I return to what might be called 'the socio-economic', which was the starting point for my historical analysis of the dairy industry, but I approach it rather differently here in the light of the foregoing analyses, with an emphasis not upon the social relations of milk as a commodity as such, but upon the ontological politics and contested knowledge-practices underpinning these relations. Thus I examine the development of the practice of milk recording in the British dairy industry in the early decades of the twentieth century, focussing upon the explication of recording as a process of rationalization, in which the formal measurability of milk yields acted as a powerful mechanism for the transformation of dairy production. In tracing this process, I outline the origins of milk recording and explore the 'social relations' which it articulated, the nature and extent of resistance to the practice, and how it underpinned the wider restructuring of the industry. I want to go beyond this crucial but limited historical-sociological account, however, by arguing that within the technical, social, and political processes of rationalization, another less tangible kind of historical process was at work, involving a reconfiguration of human–nonhuman relations. My hypothesis is that the rationalization process carried within itself a redefinition of human–nonhuman relations characteristic of the humanist work of purification.[1]

Without revisiting my theoretical and historical arguments in detail, it is perhaps worth briefly recapping here the main points of my general approach and my conception of humanist discourse: My central contention is that rather than viewing 'culture' ontologically, as the given condition of human beings and the trans-historical precondition of history, 'culture' must instead be understood genealogically as a product of history, as historically constituted. Rather than an ontological domain, 'culture' is what emerges from a set of historical practices inscribing what it is to be human. It isn't simply that culture transcends the boundaries of the human realm, but that 'culture' is an effect of such boundaries. At the same time the notion of 'culture' actively performs the work of dividing the world into human and nonhuman; it is both the mechanism and the consequence of that divide. 'Culture' then is the historical product of what I call the discourse and practice of humanism, the heart of which is an ontological division of the world into human and nonhuman spheres, and the belief that the human domain is of a fundamentally different order from the nonhuman, a belief which at the deepest level is theological; this is the condition of possibility of the dualism of 'nature' and 'culture' at the ontological heart of modernity.

This modern ontology is especially vulnerable to disordering or dissolution in the face of lived relations and encounters between human and nonhuman species, which threaten to materially undermine its structure. Nonhuman animals straddle the boundary of the human domain; their very existence continually suggests the contingency of the culture-nature divide, as they refuse to belong to either category exclusively; for this reason animals are highly problematic for the coherence of this bifurcation. To remain viable, therefore, humanism must perpetually manage the ontological threat posed by nonhuman animals; it must suppress the existential testimony of animals in order to protect the anthropocentric purity of the domain of 'culture'. Having said this, I don't want to give the impression that humanism is some kind of floating discourse, because it is very real in a material sense. Far from being a disembodied 'cultural logic', humanism needs to be grasped in terms of its *modus operandi*, or in other words how the discourse is made actual through sets of material practices and in actual history. Thus 'culture' may be historically constituted, but it must also be reproduced perpetually; it is not only a historical form of life, but is an activity and a constant process. Its reproduction demands a technical and material labour, a work of purification which involves restructuring human–nonhuman relations until they conform to the dualist categories of humanist ontology.

In this chapter, I explore the process of rationalization, via milk recording in particular, as a disciplinary mechanism through which this is achieved. My contention is that rationalization can accomplish this work because it is a potent force for reshaping material practice in accordance with the logical relations between abstract concepts; it pivots upon the ontological correction of the actual. In this way, milk recording constitutes a humanist technology, though I am not suggesting that this is all that recording amounts to or that rationalization processes do not have other objectives and consequences, such as the pursuit of economic efficiency. On the contrary, my argument is that humanist purification works within and through these other rationales and motivations.

Recorded yields: origins and resistance

Prior to the initiation of the National Milk Recording Scheme in 1914, though a small number of farmers had taken it upon themselves to keep detailed private records of their yields with the aim of improving their feeding methods, the practice was far from widespread. The *Journal of the Royal Agricultural Society* contains occasional references to the prodigious milk yields of individual cows from around 1880 onwards, but these invariably belonged to a few exceptional herds in the hands of the wealthiest and most 'progressive' owners.[2] At the Dairy Show of 1885 a paper on 'Milk Registers' aroused much interest, and two years later a Dairy Farm Records Competition was organized by the Society, with a medal being offered for 'the best kept and most practically useful record';[3] an article by the winner was published in the *Journal of the British Dairy Farmers' Association* in 1889.[4] In the following years, numerous articles advocating the keeping of milk records appeared in the journals of the Royal Agricultural Society and the British Dairy Farmers' Association.[5] Yet despite the efforts of these organizations to stimulate interest in milk recording, together with the promotional work of several Agricultural Colleges and County Councils, the majority of dairy farmers were at first not persuaded that recording was worth the time and labour involved, and for a time it remained the isolated practice of a few.[6]

There was, however, considerable geographical variation in the willingness of British farmers to experiment with milk recording, and it was in Scotland that the first attempt at systematic recording, carried out according to a definite scheme and under the auspices of a society, was made by the Highland and Agricultural Society in 1903. At first the scheme was conceived mainly as an aid to the improvement of the local Ayrshire breed, rather than as a means to systematically increase dairy output.[7] It involved the formation of local milk recording societies which would secure the services of a dairy expert to weigh, test and record the yields of the cows of the participating farmers. Towards the end of 1907 this work was transferred to the Ayrshire Cattle Records Committee, which in 1909 became the Scottish Milk Records Committee under which the scheme was extended.[8] In 1912, an annual grant of £900 was made available by the Development Fund for the supervision and operation of the scheme. This provided a model of centrally funded milk recording which began to be implemented in England and Wales two years later.

Indeed, apart from the work of a few pioneers, the progress of independent milk recording in England and Wales lagged behind developments north of the border. It was in an attempt to reverse this that the 1914 National Milk Recording Scheme was launched on the Scottish model by the Ministry of Agriculture. This provided for the payment of grants which were designed to fund part of the cost of setting up and running milk recording societies, and to support the work of the existing societies, which for several years had been urging the adoption of recording on a wider scale. It was also an attempt to widen the appeal of milk recording beyond the elite circle of wealthy farmers who were within reach of one of the many cups and trophies offered in milk yield competitions:

> The hundredweights of cups floating around the country never brought a dozen members into all the societies in the kingdom. These cups are earmarked for super herds, well managed. Circumstances compel the owners of such herds to record officially, thus making their figures reliable and acceptable to a herd book. The grants are not received from the Treasury to encourage the dozen or so herds at the top of a milk recording society in the milk yield. These owners can well take care of themselves. The official grants have the object of stimulating the average dairy farmer's interest, which will develop into an endeavour to increase herd output.
>
> (Shanks, 1930, p. 69)

Thus the Scheme marked a departure from what was a localized and exclusive honours system based upon competition for symbols of status and prestige, towards a state-sponsored programme of industrial incentives designed to increase output nationally, rather than to assist in the further development of a handful of already outstanding herds.[9] This reflected the changing organizational structure of the industry at that time, which was becoming at once more homogeneous in character and more meticulous in its level of regulation, in what was a panoptical double movement. On the one hand, faced with the threat of national food scarcity due to the looming war with Germany, the government was taking a more active role in the planning of dairy production, which was increasingly managed and organized in the aggregate, as an element of *national* agricultural production. At the same time, the regulatory regime which oversaw the industry became increasingly concerned with the micro-management of dairy farming practice, with administrative mechanisms and disciplinary technologies being set in place to ensure as far as possible its strict conformation in all conceivable details.[10] In practice, this process did not sweep all before it, however, and despite the introduction of grants the trophy system persisted, and continued to be viewed by many as an impediment to the uptake of milk recording by the average farmer:

> England is the only country where cup competition exists along with milk recording. As we gain experience of the movement, the more convinced I am that far too much energy and time of the committee and staff are directed to the consideration and distribution of these trophies. These competitions cannot induce an ordinary farmer to join a milk recording society.
>
> (Shanks, 1930, p. 70)

This may well have been the case, but there was another dimension to the relatively slow – albeit ultimately decisive – early progress of milk recording, rooted not in the tenacity of the traditional trophy system, against which the National Scheme was intended to prevail, but in a new form of social antagonism being brought into existence by the very apparatus which was to administer the Scheme.

After an initial period in which the National Scheme had little impact, in large part due to the war, the membership of official recording societies then began to grow rapidly for a number of years. While in 1916–17 only 12,950 cows were tested, in 1921–22 about 95,000 were under test.[11] Such exponential growth was short-lived, however, and in 1924 the acceleration in membership peaked and began to go into reverse, with the rate of growth of membership falling in each year from 1924 to 1929. In parts of the country some societies even found it hard to retain their existing levels of membership.[12] In trying to account for this, it is worth noting that there were sharp regional differences, with membership of societies overwhelmingly concentrated in the South-east of England where much dairy farming was relatively new, having been converted from arable farming during the agricultural crises of the 1870s.[13] In the West Counties, in contrast, which had long been the dairy farming centres of the country, membership of record societies was so low as to be almost negligible, with Cheshire and Lancashire, for example, each having a number of recorded cows under 2 per cent of the total cow population of the county, compared with 30 per cent and 25 per cent in Hertford and Berkshire respectively.[14]

This raises the question as to whether the reluctance of many farmers to take up milk recording may have been not simply the result of intransigence or ignorance on their part, as the proponents of recording tended to believe, and as official accounts therefore suggest, but rather a form of stoic or passive resistance to the practice and to the revolution in dairy farming which it represented. Thus, in areas where many farmers had been forced to convert to dairying under pressure of economic hardship in the face of foreign competition, hence where dairying had no long history but was conceived from the beginning as a capitalist enterprise, there was a significant degree of enthusiasm for milk recording. Whereas in areas where dairying was a long-established way of life, milk recording was viewed as yet another incursion into the working life of the farmer by the growing strata of officials and 'experts' who increasingly oversaw the industry:

> The keeping of a record of the milk yield is a businesslike way of working, and the farmer who does not look upon his work merely as an occupation, but as a business, and does everything in a business way, is likely to be the most economical and, financially, most successful farmer.
> (Lawson, 1913, p. 63)

This is given further credence by the fact that an increasingly large number of farmers, while recognizing the usefulness of milk records, chose to eschew the state-authorized schemes of the Milk Recording Societies, which were obliged to follow the stipulations of the Ministry of Agriculture for official recording, opting instead to undertake their own milk recording practices tailored to their own perceived needs. It was equally the case of course that in the areas where dairy farming was most established, the existing practices were regarded as working adequately and thus milk recording was deemed unnecessary;[15] however, such an explanation does not weaken the case for positing a dynamic of resistance but is better seen as its manifestation. For it was precisely in the tension between tacit practical knowledge and formal-rational scientific knowledge that the conflict between working farmers and bureaucratic experts was in large part played out within the discourse of modernization, and the relative efficacy of these different ways of knowing was intrinsic to this struggle. Equally, it was partly through such contestation as to the comparative effectiveness of these knowledges that the conflict between the two groups was actually constituted. To view the practicability of contending forms of knowledge as a matter separate from the social conflicts of their proponents is to dislocate 'the social' from its embeddedness in material processes and cast it as a sphere of pure meaning, in what is an unwarranted separation of the nature of matter and the struggles of power.

In discussing the possible reasons for what he regards as the disappointing rate of uptake of Record Society membership, especially in the West Counties, James Macintosh – a high profile dairy expert and supporter of milk recording writing in the *Journal of the British Dairy Farmers' Association* – suggests three main credible explanations. The first and second of these tend to confirm the argument that many farmers eschewed official recording because they were disinclined to submit to the forms of regulation and surveillance which it involved:

> a) That they keep private records, and do not believe that the information which official milk records supply is worth the money, time and labour spent in obtaining it.
> b) That they give general adherence to the principle of milk recording by societies, but they consider many of the regulations imposed by the Ministry of Agriculture unnecessary and irksome, and prefer not to be bound to comply with them.
> c) That they do not realise the value of milk records as a guide to profitable management.
>
> (Macintosh, 1929, p. 71)

In making the case for recording, Macintosh proceeds to dispute the validity of each of these reasons which a farmer might give for deciding not to join a recording society. His argument in relation to the first is particularly revealing:

> It is quite true that properly kept private records can supply all the information needed to detect the good and the bad cows, to enable the herd to be fed wisely, and to improve the general management, but private records cannot have the commercial value of officially checked records when it is desired to sell a good cow or her progeny, or when a herd must be disposed of.
>
> (ibid.)

Significantly, this is a complete reversal of his stated opinion eight years earlier that focussing upon the value of milk recording in the *sale* of cows, in the sphere of exchange as opposed to production, was a distraction from their intended, and ultimately more important, use in increasing productivity, centrally through the rationalization of feeding and breeding practices.[16] What such a reversal of emphasis most plausibly suggests is not that Macintosh has changed his opinion on the relative importance of milk records for the profitable sale of cattle or for increasing productivity, for if this were the case, then some reason for the change would surely be given; rather it suggests that in the intervening years *official* milk records have become an end in themselves, for reasons other than those of the productivity of individual herds. Since official records clearly offer no distinct advantage over private records in terms of herd productivity, Macintosh must grasp for any reasons available to promote membership of the official schemes. For example, although he goes on to admit that the overwhelming majority of dairy farmers are not members of breed societies, which generally included only the wealthiest farmers, he nevertheless cites as a key advantage of official recording the requirement of breed societies that yields for entry into their year books or their advanced registers have been checked by a society operating under the auspices of the Ministry of Agriculture, and that calves for entry into their supplementary and grading-up registers be earmarked by a Milk Recording Society: 'Hence, membership of a society is necessary in order to take advantage of many of the helpful schemes carried out by the Breed Societies.'[17]

The reason for this is bound up with the whole trajectory of the development of the industry in the early to mid-twentieth century, which involved increasingly close regulation, both by the state and by various sub-state political agencies. For example, part of the stipulations laid down by the Ministry of Agriculture for the official milk recording schemes, and a condition of obtaining the grant available for setting up and administering such schemes, which were calculated on a herd-by-herd basis, was that the annual return of the milk yield for the herds be supplied to the Ministry.[18] These figures were then entered into a National Register of dairy producers, the like of which had never been previously attempted. Far from milk recording then being a purely practical and economic scheme for increasing productivity, it was very much a political exercise in amassing knowledge of the population through statistics. Under the regime ushered in by milk recording every increase in productivity and private profits was matched by refinements in the technologies of power.

Official milk: situating the producer

If many among the more experienced dairy farmers were resistant to milk recording and to the forms of surveillance by experts and officials and the regulation by authorities that it entailed, this is unsurprising in light of the preceding conflicts between farmers and local government which had arisen from the imposition of the 'milk standard'. The term referred to the legally designated minimum percentage levels of the various constituents of milk, such as the proportion of fat, which was officially used to determine whether milk was 'genuine' or adulterated. The 1899 Sale of Food and Drugs Act gave the Board of Agriculture powers to regulate the minimum levels of the constituents of milk, and in 1901 the Sale of Milk Regulations laid down a standard of 3 per cent fat and 8.5 per cent solids-not-fat below which milk was presumed to be adulterated.[19] The fierce conflicts engendered by this policy highlighted a tension not merely between group interests but, in and through these, between contradictory visions of nature and antithetical notions of what constituted 'pure' milk.

For working farmers, milk was an inherently variable natural substance, subject to a plethora of subtle natural rhythms and influences beyond precise control, as the often marked difference in the butterfat content of the produce of a single cow milked at different times of the day served amply to demonstrate. Hence any milk standard that stipulated rigid minimum constituents as definitive indicators of adulteration would almost certainly lead to widespread accusations of malpractice where there had been none.[20] While many producers were themselves keen to stamp out adulteration and its devastating impact upon the public perception of milk, they were deeply wary of the notion that adulteration could be detected simply by an analysis of its constituents; for them the truth of production did not lie in the composition of the product. For the food inspectors and local sanitary officials charged with detecting and prosecuting adulteration, however, a legally defined milk standard was an indispensable weapon in the emerging public health arsenal. Without it, the detection of adulteration could not hope to keep pace with the rapid expansion of the industry, as the difficulties of literally proving adulteration in any given case were almost insurmountable.[21] Faced with this dilemma, the authorities reversed the burden of proof, obliging the accused to prove their innocence when charged with adulteration, and effectively using a fixed legal standard to criminalize vast numbers of producers throughout the industry:

> A public analyst examining a sample, and finding it below the standard, usually states that a certain percentage of butter-fat or cream has been abstracted, or a certain percentage of water has been added. He has no

> kind of right whatever to make such a statement, as it is at best a mere matter of 'presumption'. The sample may have been drawn from perfectly healthy cows, properly housed, groomed and fed. It may show a deficiency of fat or non-fatty solids, and yet be genuine. It appears likely that the fact of the impossibility of 'proving' adulteration in the majority of milk cases is responsible for the departure – in cases where milk is concerned – from that good old axiom of English law, that a man must be considered innocent until he is proved to be guilty. The milk purveyor, whether farmer or dealer, prosecuted under the Food and Drugs Acts is held to be guilty until he can prove his innocence.[22]
>
> (Sadler, 1907, p. 24)

The ensuing conflict between farmers and local public health officials was rooted in the contradiction between the natural irregularity of the material world and modernity's political and economic demand for standardization. The will of monopoly capital and the modern state to control and regulate production within a territory clashed with the intrinsic 'unreliability' of cows as nonhuman producers.[23] It was essential that the product conform to a fixed standard if its mass consumption was to be effectively organized and regulated, and yet the irreducible corporeality of the substance in question, its stubborn nonhumanity and its dependence upon interspecies relations of production, all militated against any smooth standardization. The method by which this material resistance was overcome is of significance both for the overt process of rationalization within the industry and for the operationalization of humanist discourse which took place within and through this rationalization. By the fixing of the 'milk standard' the 'unreliability' of the cows themselves was effectively negated and the farmer made wholly responsible for any and all irregularity in the product. Thus the role of cows as co-producers was effaced through the imposition of bureaucratic and judicial universalism upon the product. Milk was thereby not merely standardized but at the same time *humanized*; the variability which attested to its non-human 'nature' could not perhaps be obliterated materially, at least not yet, but it could certainly be erased officially and legally. Hence farmers became *de jure* the sole producers of milk, cows being merely the organic machines which they utilized for this production.

Through this process the circle of human production, of 'society', was redrawn in such a way as to exclude nonhumans the more consistently from its domain. Relations between two species were turned into relations between 'nature' on the one hand and 'society' on the other, there a world of quasi-objects and here a world of subjects. Thus the sphere of 'culture' was again secured and reproduced in its anthropocentric purity, its secular divinity. Such a strategy could only ever represent a partial solution, however, while milk *de facto* continued to materially defy the categories through which it was economically, legally and ontologically managed. Indeed, it was the recalcitrance of milk itself, the obdurate resistance of its very materiality, which underpinned and fuelled the struggle between dairy farmers and health inspectors, in many cases ending in the courts; thus these courts became arenas for the enforcement of humanist ontology. Just as the courts of the Middle Ages had enforced that epoch's vision of a hierarchical yet continuous relation of humans to nonhumans, by prosecuting individual animals for crimes against their human owners, now these courts enforced modernity's radically different vision, by prosecuting farmers for the refusal of their animals to conform to human rules.[24] This inscribed the incommensurability of human and nonhuman existence, the essential duality of 'culture' and 'nature', and the absolute immanence of 'culture'. In the feudal courts animals were situated, albeit ambiguously, *within* the circle of justice, of values and culture, as morally responsible beings; but in these later prosecutions of dairy farmers cows were located unambiguously *outside* of the circle of (human) justice. Thus the political apparatus of legal enforcement was called upon to define not merely what was sanctioned by the law and what was proscribed, but at the same time what was endowed with the capacity for meaningful 'agency' and what was not.

If the farmer was to be held responsible and charged with adulteration whenever his milk was found to be 'substandard', the corollary was that the cows themselves had to be seen as incapable of variability or temperamentality; they had to be conceived as complex organic objects passively reacting in predictable ways to various external influences, like the clockwork creatures of Descartes' mechanistic materialism. This involved a double politicization, at once social and ontological, with each dimension underpinning the other: On the one hand, farmers were subjected to the scrutiny of urban milk and sanitary inspectors and the judgment of the courts; but within and through *this* process animals were regulated both discursively and materially and their absolute otherness in relation to humans both reinforced and made enforceable, which in turn reproduced the ontology upon which the authority of the courts was ultimately founded. Thus the criminalization of the human producer was inseparable from the purification of the animal producer, its complete removal from the sphere of 'culture', of 'agency', the expunging of its social 'soul'. As a vehicle for purification-rationalization, the milk standard was nevertheless imperfect, for it was fated to battle constantly against the 'irrationality' of the cows themselves, mediated by the composition of the milk, which persisted in varying wildly without regard for the official standards. In order to suppress this existential testimony of nonhumans, the process of standardization had to go beyond bureaucratic and judicial categories and enter into the very corporeality of the animals; the image of cows as constant and reliable machines had to be made materially actual.

This parallels the more general development of humanist discourse in the period of interest, which in order to contain the ontological insecurities and contradictions arising from the incipient transformations of modernity, was compelled to enter increasingly into scientific and technical knowledge-practices and to assume an increasingly material form, to invest itself within an apparatus both material and epistemological, hence to become real or 'objective' in this sense. If even unadulterated milk frequently fell short of the required standard, then its production had to be brought under rational control until such inconsistency could be extinguished; modernity demanded nothing less. Thus the milk standard was tied inextricably to milk recording, which was in turn the first step towards the real standardization of the cows through rational breeding. This shift in species relations was mediated once again through the 'agency' of the milk; just as tuberculous milk had entered into modernity and destabilized the ontology of 'culture' at its heart, forcing humanism into a restructuring centred upon its material operationalization, so now modern rationalization entered into milk production and precipitated the material transformation of the producer species. Thus the frontier of 'culture' was extended further into 'nature', keeping its theological heartlands safe. The species encounter embodied in dairy milk was thereby sanitized for human consumption and cleansed of the traces of its human–nonhuman hybridity. Standardization of the product then was not only economically necessary but was integral to its inscription as 'cultural', to its humanization, and to the removal of its animal 'nature', a process which ultimately required the real standardization of the animal producer. Milk recording was a technical lever by which these overlapping rationalizations were achieved.

The uses of recording I: scientific feeding

In practical terms milk records were used at two principal levels: at the national level they were utilized by the Ministry of Agriculture as a source of statistical knowledge with which to inform dairy policy; but most importantly they were used by farmers as a source of information with which to guide everyday decisions on dairy practice. Their effectiveness lay in the fact that, once in use, the records introduced a quantitative rationality into farming practice which was experienced by farmers as external necessity; the internal logic of the method was able to dictate farming practice more effectively than any central control could ever have done. Records effectively took dairy farming as a complex and integrated way of life and broke it down into a finite number of definite variables; thus the forms of tacit and practical knowledge which had been at the heart of traditional farm management were increasingly displaced by statistical monitoring and rational manipulation of the factors affecting milk output. Although such factors as the time and frequency of milking, the type of housing and the age of first calving were also involved, the main variables in question centred upon feeding and breeding, each of which I will discuss in some detail.

According to its proponents, feeding practices were the aspect of dairy management which could be most immediately affected by recording, since changes in feeding methods would produce a dramatic impact upon milk yield in a fairly short time-frame. Whereas the benefits which could be secured through selective breeding for higher yields took several generations of animals and therefore many years before they became apparent, changes in feeding regimes in the light of recorded yields could be based upon a relatively short period of experimentation, and the results felt almost immediately. The ultimate aim as far as the government was concerned was to determine as quickly as possible the single most efficient system of feeding which would produce the highest national output, with a view to implementing this scientifically optimal system across the board. Towards this end, the Ministry of Agriculture in 1924 appointed a Departmental Committee to advise as to 'a uniform scheme of advice on the rationing of dairy cows for general adoption throughout the country'.[25] This attempt to move towards a uniform national system of rationing was typical of the trajectory of modernization in the industry, which involved seeking to realize a model of standardized, high volume, low cost, mass production, for a mass market which was even then being created. Demand and supply, consumption and production alike were to be controlled and regulated in the 'national interest'. In practice, however, as in so much of dairy modernization, the movement towards a uniform system, though it did occur, was rather patchy, faltering and uneven. Thus, even in 1930, there was still no consensus as to what feeding regime was the most productive, and there were persistent calls for more systematic research to be carried out on the matter.[26]

One aspect of feeding in which there was a general shift, precipitated to a significant extent by milk recording, was in the move towards increasing use of concentrated foods. These were heavily promoted by the recording societies, most of which had entered into arrangements with the Agricultural College or the University Agricultural Department in their area, or with the County Agricultural Organizer, to provide their members with advice on the selection of rations and the purchase of concentrates.[27] This was a significant turnaround, as the conventional wisdom for decades had been that any article of diet other than grass was at best a supplement or a useful but inferior substitute, to be used only when the climate, the season, the urban location, or the demands of field management for arable farming, meant that grazing was either out of the question or temporarily impossible.[28] The assumption was that grazing was superior because it was the more 'natural' method of feeding, and this appeared to be borne out by the generally poorer quality of milk which was believed to be produced by the town cowhouses that had survived from the period before the cattle plague, which were largely fed on a diet of grains. Town milk was often found to be watery in comparison with that of rural dairy farms and this was attributed to the 'unnatural' diet of town herds, although the extent to which this impression may have been due to the widespread practice of watering down by urban milk sellers is impossible to determine. With the new popularity of palm kernel cake, bean meal and brewer's grains, however, the age-old belief that a 'natural' diet was inherently superior was finally shaken; the faith in nature that had guided farmers' decisions on feeding practices had begun to be displaced by a faith in science and 'progress', which is to say, a faith in modernity.

This was due in part to advances which were being made in veterinary nutritional science, enabling the production of concentrates of far higher quality than anything which had previously been available. It was also due to the new interest which farmers were taking in how the exact quantities and proportions of the different constituents of a feed could be varied in order to influence the milk-producing capacity of their cattle. Such interest was engendered to a significant extent by milk recording, which showed for the first time in a quantifiable form that close attention to the details of bovine nutrition could save farmers large sums of money.[29] Thus those who could afford to do so were increasingly buying pre-mixed commercial feeds in the form of concentrates in the hope of augmenting the yields of their herds.[30] This was also a consequence of aggressive advertising, for throughout the trade papers and agricultural journals of the time bold claims were repeatedly being made concerning the milk-stimulating capacities of these concentrates. Hence this period marked the large-scale uptake of systematic rational feeding based upon both milk recording and the scientific research into bovine nutrition that was being undertaken in the newly founded dairy research centres, such as the National Institute for Research in Dairying, which was set up in Reading in 1912.

Such research was not in itself entirely novel; the chemistry of animal nutrition had been studied in increasing detail since the 1840s, with the composition and nutritional value of a variety of feeds being analysed and assessed, and the results printed in the agricultural periodicals, trade papers and farming textbooks, particularly from the 1870s onwards.[31] Their influence was limited, however, for many farmers were sceptical of the real-world applicability of these experiments, preferring to rely upon their own forms of practical empiricism.[32] With the development of milk recording though, the impact of such research was magnified exponentially, for even the most sceptical farmers could now see for themselves the relationship between nutrition and milk yield in a quantifiable and objective form.[33] Although there had always been a minority of farmers who took an interest in the results of the various experiments and who might be prepared to adapt their practice accordingly, these were very much the exception and the average farmer had stuck fairly rigidly to his established feeding methods.[34] By the 1920s though, it was a matter of the utmost concern to every farmer whether his methods were as efficient, economical, and ultimately as profitable as could be achieved in the light of the scientific evidence which was becoming ever more widely available.[35] Science and dairy production, which had previously been semi-detached, with veterinary nutrition on the one hand and practical dairy farming on the other, were suddenly thrown into an indissoluble union.

Milk recording then was a central mechanism for the transformations that were taking place during the second and third decades of the twentieth century in the economic, political and epistemological relationship between an expanding infrastructure of scientific dairy research and education on the one hand and dairy farming practice on the other; hence between technical experts and officials as the bearers of the new knowledge and working farmers as representatives of traditional methods. Farmers increasingly found that they could not afford to maintain their customary indifference towards institutional dairy research in what was an ever more knowledge-intensive industry. Their growing dependence upon the scientific research being carried out by specialist researchers removed from the practical knowledge involved in farm work significantly compromised their autonomy and their ability to resist the encroachment of the regulatory authorities, who could increasingly deploy scientific and specialist knowledge in extending their control; thus farmers began to become dilettantes in their own work. Early attempts at regulation of dairy production had been severely limited by the relative ignorance of the various inspectors, who were invariably charged with overseeing what they understood only in terms of legislative strictures. Indeed, the agricultural journals of the late nineteenth century are full of biting and witty commentary by working farmers on the ignorance of farm practice characteristic of those administering this or that aspect of dairy production:

> The reviewers have been, in most cases, profoundly ignorant of the subject, so that their conclusions are usually contradicted by facts, while their suggestions are bewildering, not only from their protean variety, but from their absurd impracticability. Unfortunately, our educational system seems 'so careful of the type' of half-fledged scientists that for one who succumbs to the logic of facts a dozen start into being, each one with a more advanced – shall I say wilder? – theory than his predecessor.
>
> (Marshall, 1906, p. 17)

> The majority of them have learnt their chemistry, and bacteriology in a laboratory, but do not know a cow from a horse except by the horns, and have still to learn the practical application of their knowledge.
>
> (McConnell, 1917, p. 89)

It is telling that by the 1920s such comments have largely disappeared. In their place is a feverish concern with the minutiae of dairy science, and the character of a significant part of the writing in the journals by now involves 'experts' explaining the essentials of their field as pertains to some or other aspect of dairy production to a presumed audience of lay-person farmers. Moreover, the earlier journals were practical through and through, centring upon the recommendation of minor reforms rooted in experience and best practice and usually with a tone of bucolic common sense; but by the twentieth century specialist scientific knowledge had become, if not an end in itself, then certainly the driver of practice rather than merely its guide, and the tone now is more abstract, definitive and authoritative. The dynamic of gradual improvement which had characterized dairy practice for centuries was ruptured by a dynamic of urgent transformation; the hysteria of modernization became evident. A harbinger of this shift was the debate over the communicability to humans of bovine tuberculosis, which very often pitted microbiologists against farmers;[36] hence the struggle against milk-borne disease was the breaking ground in which the authority of the knowledge of the laboratory over that of the field was finally stamped beyond doubt.

Allied to this transformation and operating both within and through it, the spread of milk recording also impacted upon the ontological positioning and classification of the animal in relation to 'society', to 'social' production, and thus in relation to humans. The development of rational feeding inscribed the animal body as a processor of nutrients, as a 'black box' which performed the alchemy of turning plant foods into milk. Of course the cow had long executed this key function for dairy farmers, but now for the first time all other considerations were formally excluded from what became an instrumentally rational analysis of input and output ratios. Rather than conceiving of the cow as a partner in a naturally unequal relationship, as was typical of the farmer's traditional view of the role of the dairy cow, the animal was now reduced by the science of veterinary nutrition to an organic object, a machine which manufactured the raw material of grains and concentrates into the desired consumer product. The feeding stuffs were both the fuel for this machine and its primary material, and there was a preoccupation with what precise mixture of these fuels and raw materials would produce the greatest quantity of the product, as measured by milk records. Attempts were made to calculate what quantities of various different foods should be fed at a time in relation to a cow's bodyweight in order to obtain the maximum milk yield while maintaining the existing bodyweight of the animal;[37] the formula could then be adapted accordingly when it was desirable to 'fatten up' the animal for slaughter. Thus milk yield recording set in motion a self-reinforcing circle of economic valorization and analytical reduction, in which yield statistics provided both the means and the motive for a rationalization of feeding practices, which in turn confirmed in its results the ultimate value of recording. The processing of plant foods into milk, of 'nature' into 'culture', became no longer one aspect of the cow's life-activity, which happened to be of use to its human owners; instead this function was increasingly the sole purpose of the cow's existence.

Perhaps the starkest example of this logic at work was to be found in the practice of forced feeding, which enjoyed considerable support at this time and involved the systematic overfeeding of cows over a more or less extended period in an attempt to stimulate greater milk production. It should be said, however, that this was never universally accepted as the standard method of feeding and there were always farmers who opposed it, most often on the grounds that it would quickly destroy the health of the animal and so be uneconomical in the longer term:

> Forced feeding will increase the milk yield of dairy cows of the right type, but the question as to the desirability of adopting such a course is a

> debatable point. One can understand how a cow being tuned up to give the maximum for a test such as the Dairy Show demands, but such handling could not continue for any length of time.
>
> (Shanks, 1930, p. 71)

There were also significant differences between breeds in the extent to which the excess food consumed through forced feeding would be converted into milk. Of the various breeds, the Friesian was believed to be most amenable to successful manipulation in this manner, and this was a critical reason why this animal began to replace the established local breeds and was increasingly accepted as the standard dairy breed across much of Britain.[38]

Balancing mechanisms: the return of the repressed

If there were concerns in some quarters that the brutal rationality of forced feeding would soon destroy the health of an animal, there were similar concerns in other areas that rationalization would give rise to feedback mechanisms which would ultimately undermine the productivity which was its objective. It was around this time, for example, that farmers in Cheshire and Somerset began to notice that despite the prospect of rapidly increasing yields made possible by milk recording and scientific feeding, there appeared to be a countervailing tendency for milk yields to diminish over time.[39] The explanation for this was that the quality of the soil was being slowly impoverished by the continual abstraction of its nutritional and phosphate content in the form of milk. This fact powerfully re-established the connection between land and milk, hence between farming as an arduous process of scratching a living out of the land, and the dairy industry which many farmers had entered precisely as an escape from the privations of arable farming and its perpetual struggle with the land. Just at the moment when the modernization of dairy farming was accelerating and its subordination of nature began to seem unstoppable, with the factors involved in dairy production being simultaneously complexified at the level of scientific analysis and homogenized at the level of practice, and just at the moment when the cow as a result began to be transformed into an industrial processor of nutrients, suddenly an older conception of nature threatened once again to reassert itself; this was a view in which 'progress' was finite, circumscribed and vulnerable to natural 'agency'.

In order to contain this re-incursion of 'nature' into the advance of 'culture', efforts began to determine how much land was required for each cow if improved yields were to be sustainable over time. In addition, agricultural research began to turn its attention to the question of how soil phosphates depleted by intensive dairy farming could be replaced rapidly by artificial methods; thus a practice which would soon culminate in the general and widespread use of chemical fertilizers was born. These strategic responses were complementary; for once land requirements in relation to herd size were determined with some accuracy, then the rationalization process which was centrally responsible for the problem could be turned against it and deployed in its solution. The hope was that by studying milk yield statistics over time in relation to land area it would be possible to measure the effect of fertilizers in terms of reduced land requirements and thus to gradually reduce the land required per cow through the combined use of statistics and chemistry. But in practice the victory of rationalization over its own unintended consequences would never be complete, and depleted soil quality due to intensive milk production would remain a major problem for dairy farmers throughout the twentieth century, necessitating regular advances in scientific chemistry and steady increases in the use of chemical fertilizers in order to stay one step ahead of the creeping threat of diminishing yields.

Another area in which careful regulation was necessary in order to circumvent the balancing mechanisms characteristic of natural 'agency' was in the matter of milk 'quality', which since the enforcement of the 1901 Sale of Milk Regulations had been legally defined as its butterfat content. It was soon observed that increases in milk yield secured through scientific feeding, and particularly through forced feeding or the overfeeding of certain feeding stuffs which served to stimulate milk production, such as bran, were very often matched by a reduction in the butterfat content of the milk. In the simplest terms, increased quantity was in many cases bought at the cost of reduced quality. Indeed, in Scotland the problem of this inverse relationship was identified from the earliest days of milk recording, hence Scottish milk recording always incorporated tests of butterfat content, variations in which could then be studied in relation to increasing yields in order to identify and determine how to tackle any trend towards steadily diminishing quality. Once again a problem arising from the intensification of dairying facilitated in large part by rational measurement was reckoned back into the rationalization process and contained by being subjected to measurement itself. In England and Wales, however, despite the recommendations of early supporters of milk recording, the Milk Recording Scheme inaugurated by the Ministry of Agriculture in 1914 differed from those in practice in Scotland, Ireland and Denmark in that it did not provide for systematic and regular testing of samples of milk from each cow for their milk fat content. In these countries the testing of the quality of the milk from each cow was an integral part of the milk recording scheme; whereas in England and Wales the taking and testing of samples were optional at best, and societies were only prepared to carry out this work at the express request of members, and then usually only for bulk samples or for a few individual cows.[40] Furthermore, although the method by which samples should be taken by recorders was set out in detail by the Ministry of Agriculture, no specific method of testing for quality was established and no directions were given as to how the average percentages of fat for a day or for a lactation period should be calculated. There are two likely reasons for this.

It may have been that the regulatory emphasis which had been placed on the moment of consumption by the 1901 Sale of Milk Regulations was deemed sufficient, so that the establishment of a fixed legal minimum proportion of butterfat content for milk exposed for sale in England and Wales was regarded as adequate for the purposes of ensuring consistency in milk quality. This would be congruent with the increasing separation of the moments of production and consumption that characterized the developmental trajectory of the industry, and with the preference for indirect regulation of producers via 'public health' legislation targeting the sphere of retail and consumption which characterized dairy regulatory strategy until the Milk Marketing Acts of the early 1930s. This could also be observed in the administrative response to tuberculosis, in which the sale of 'clean' milk free from tuberculous bacteria to 'the public' was carefully disconnected from the question of its production by animals who were themselves free from tuberculosis; public health and agriculture were thereby kept separate and pasteurization was the technological expression of this separation. Alternatively, it may have been that whereas in Scotland the clerical work of recording had been taken out of the hands of farmers and carried out wholly by specially trained recorders from the very beginning, in England and Wales many farmers undertook the work of recording themselves, and so quality testing was considered an additional burden which might discourage the uptake of recording.[41] Indeed, the actual keeping of records entailed a considerable bureaucratic workload and this was often identified as a key reason for the relatively slow progress of recording in England and Wales, and for the necessity of the financial incentives provided by the government under the 1914 scheme. To increase this workload above the absolute minimum through the introduction of this very scheme may have seemed a contradiction too far, given that its main objective was to encourage the further uptake of recording.

Whatever the reasons, the result was that in England and Wales the question of the improvement of milk yields was to a significant extent dissociated from that of the quality of the milk. This impacted upon feeding practices, since it fostered the continuation of feeding methods which were detrimental to milk quality over time, but it also had serious implications for breeding practices:

> By concentrating on yields and neglecting the quality of milk in the selection of cows and purchase of bulls, a farmer may be steadily reducing the average quality of the milk of his herd, and to a certain extent also of the breed to which it belongs. Thus a milk record scheme which concentrates attention on yields only may do a considerable amount of harm, by unconsciously lowering the average quality of the milk.
>
> (Macintosh, 1921, p. 39)

This was a cruel paradox given that breeding was the aspect of dairy farming practice for which milk recording had seemed to promise the most:

> While milk recording provides almost immediate means of improving the milk yield and the feeding of a dairy herd, probably its biggest contribution to the dairy farming industry lies in its present and potential value as an aid to the breeding of dairy cattle for milk production.
>
> (ibid., p. 40)

The rationalization and intensification of dairy production through milk yield recording were by no means a straightforward, unilinear, or uncontested process then, but rather were riddled with contradictions and unintended consequences, arising both from their own logic and from the inertia and resistance of the material forms which were their object. Central to these problems was the conflict between quality and quantity, in which negative feedback mechanisms tended to hollow out and undermine the process of quantification. Indeed, as rationalization processes laboured to produce an ever more immanent cultural sphere of social production, a purely human domain of reason and progress divorced from nature, it faced an increasingly powerful return of the repressed. This in turn created the need for an ever more comprehensive circle of scientific control and intervention, taking account of all known variables, since any gap in the circle and any variable left uncontrolled would provide a point of re-entry into the cycle for the counter-balancing mechanisms characteristic of natural 'agency'. This throws some light upon the necessary laboriousness and ceaselessness of humanist technologies in the face of the recalcitrance and stubbornness of organic matter, which is always tending to return to its own rhythms, whether by death, disease, or whatever form of decay.

The uses of recording II: rational selection

The selection of animals for breeding purposes was the pivotal and defining use of milk records which more than anything else rendered them valuable. Selective breeding in itself of course was nothing new; it had been going on for centuries albeit in a somewhat haphazard manner, and by around 1800 the slow accumulation of its results had led to the emergence of the distinctive characters of the modern breeds, which for the first time began to be carefully noted in farming textbooks.[42] Prior to the spread of milk recording, however, farmers lacked any rational basis for selection; instead there was an intricate practical lore which had been established through generations of farming experience. This centred upon the deep-rooted belief that an animal's milk-producing capacity was correlated with certain features of its appearance, with particular emphasis upon the parts close to the milk-producing glands.[43] Thus farmers would look for cows conforming as closely as possible in all details to the following description:

> Head small, long and narrow towards the muzzle. Horns small, clear and bent and placed at a considerable distance from each other. Eyes not

> large, but brisk and lively. Neck slender and long, tapering towards the head and a little loose skin below. Shoulders and forequarters light and thin. Hindquarters large and broad. Back straight and joints slack and open. Carcass deep in the rib. Tail small and long, reaching to the heels. Legs small and short, with firm joints. Udder square, but a little oblong, stretching forward, thin skinned and capacious, but not low hung. Teats small pointing outwards, and at a considerable distance from each other. Milk veins capacious and prominent. Skin loose, thin and soft like a glove. Hair soft, short and woolly. General figure, handsome and well proportioned.
>
> (Harley, 1829, pp. 29–30)

Though this was supposed to be broadly applicable to all breeds, it was usually influenced by the breed with which the describer was most familiar.[44] Indeed, prior to milk recording, though farmers were emphatic about their choice of a cow and attempted to select the best milk-producing animals, they were geographically restricted in their choice and most were content to stick to the local breed.[45] Moreover, in marked contrast to the attention that was given to the selection of cows, very little consideration was given to the choice of a bull; in most cases any bull was used.[46] The exceptions were those who bred pedigree animals for competition, who had realized earlier than most that the choice of a bull was just as important as the choice of a cow in determining the qualities of the progeny.

Although those who could afford to breed animals for competition were separated by a social and economic gulf from the vast majority of dairy farmers, the knowledge of breeding which arose from the manufacture of pedigree did influence the breeding practices of ordinary dairy farmers to a certain extent. Consequently, though many were still prepared to accept any bull, by the 1870s, a minority were already beginning to look for an animal that came from a good milking family. They realized that a bull of this descent would soon improve a dairy herd if the heifer calves were reared for replacements, a distinctly modern idea.[47] What they lacked, however, was any objective basis for determining whether a bull was indeed from a family of productive milkers; in most cases they had only the word of the previous owner, who was usually attempting to sell the animal at the highest price and could therefore be expected to be somewhat economical with the truth. Even where a bull's history could be ascertained with some credibility, it was very often found that the animal did not necessarily pass on the desired qualities as anticipated:

> The bull may be pleasing to the eye . . . but we are up against traits which may develop in the progeny, contrary to expectations. A good dairy bull does not always produce heifers which are a credit to the herd in the way of milk yield.
>
> (Shanks, 1930, p. 74)

Some farmers allowed this to demoralize them as to the very possibility of improvement through selection and many resorted to their residual fatalism, declaring that breeding for milk was in the end a game of chance.[48] This opinion was by no means unreasonable in the circumstances; it reflected the fact that what they were engaged in was an attempt to produce very specific desirable outcomes, which were actually the result of the interaction of large numbers of complex genetic variables, through the very crude instrument of selection. That was why even a bull and a cow with excellent qualities themselves could sire heifers with disappointing milk-producing capacities.[49] Milk yield recording in and of itself did not solve this problem overnight, but it did open up a whole new plane of knowledge, with new possibilities for knowing animals as both individual productive units measured across the useful lifespan of the animal, and as pools of inheritable traits measured across multiple lifespans. Ultimately this enabled the distinction to be made between cows which were merely valuable in themselves and those which could be relied upon to transmit their valuable qualities to their heifers.[50] Thus the detailed study of milk records enabled the bodies of animals to be known and controlled in a previously unimaginable way, as invisible potentialities operating across generations.

One notable result of the use of milk yield records in selective breeding was in the belated demise of Guernon's 'escutcheon theory', which had for decades been highly regarded as the closest thing to a scientific method for the selection of dairy cows. Guernon's central assertion was that it was wholly possible to judge a cow's milk-producing capacity by what he called the 'escutcheon'. This consisted of the udder and the area above it between the thighs and hips, as viewed from the rear; particular importance was ascribed to the direction of the growth of the hairs in this area. Guernon claimed to have made comparative observations over a protracted period, until despite some aberrations he reached the conclusion that there was a definite relationship between the shapes of curls and the direction of growth of hair, and the productive capacity of a cow.[51] The theory 'became immediately popular, and was enthusiastically subscribed to by the highest contemporary authority'.[52]

While at first it is difficult to understand retrospectively why it was that such a doubtful theory was so readily embraced, it is worth remembering that until the dawn of milk recording the practice of selective breeding was heavily rooted in the belief that a cow's outer appearance corresponded in some way to its productive capacity and that successful selection depended upon the skill of the farmer in correctly reading the various signs which were a covert expression of this information. Guernon's theory simply offered a more explicitly codified form of this belief and purported to have extracted some reliable essence from the plethora of physical features that farmers were accustomed to having to try to make sense of. Thus Guernon offered exactly what was needed at the time in order to give farmers more confidence in their decisions, and the fact that it was scientifically untenable was essentially irrelevant. Until some superior paradigm displaced it, Guernon's theory served adequately to reproduce the theological notion that human reason was a higher power than nature; hence its ontological work was unaffected by its material inaccuracy. This was reinforced by the fact that, though it was seductively simple in essence, the specifics of the theory were incredibly convoluted and often contradictory, so much so that its failure to produce the desired results in any given instance could always plausibly be attributed to the failure of the farmer to properly apply the theory. So circular was the belief system that no isolated piece of evidence could dislodge it. In this light, Guernon's theory is perhaps not so very different from much of what passes for social scientific knowledge even today, and one can begin to understand why the escutcheon theory continued to be well regarded by many until well into the twentieth century.

The method of selecting milk-producing animals by their physical appearance was highly influenced by the breeding of pedigree animals for display and inspection within the 'show' system; the logic of production as yet was subsumed under the logic of the spectacle:

> There can be little doubt that the development of the show system in the latter half of the nineteenth century and the demand from other countries for animals of certain lines of breeding concentrated attention on high condition and the possession of breed characteristics which were not necessarily combined with utility and efficiency in the production of milk. Up to this stage judgement and selection of dairy stock had been almost, if not entirely, based on opinions unsupported by reliable information in the actual quantity and quality of the milk yielded by the cows. Essential data by which judgement by the eye and hand could be checked and corrected were not available.
>
> (Macintosh, 1938, p. 34)

The organizational transformation of milk recording signalled by the National Milk Recording Scheme, however, away from an honours system organized around competition trophies and inspection prizes and towards a co-ordinated national system supported by government grants, precipitated a parallel shift in breeding practices. The importance of 'show' animals steadily declined as the emphasis turned from breeding magnificent beasts who could win prizes for their owners at the high profile dairy and agricultural shows, towards breeding animals for maximum milk yield.[53] It was partly in reflection of this shift and partly to help propagate it further, that the British Dairy Farmers' Association introduced butter and milking trials at the London Dairy Show in 1879.[54] The breeding of pedigree animals for visual inspection, though enthusiastically pursued by those few who had the means, as a way of displaying and expressing both their wealth and their nobility, became an ever more exclusive, minority activity.[55] It was increasingly separate from the main developmental trajectory of the industry, which was becoming ever more squarely focussed upon productive capacity. Together with the introduction of the National Milk Recording Scheme, this was part of the wider process of bourgeoisification of the dairy industry, which is to say the gradual wresting of its political centre of gravity from the gentry, whose domination of its local and national associations was reinforced by the distribution of prizes, trophies and cups:

> The class of individuals who lifts the cups is usually fairly well represented on the committees, and it is amusing to note how they endeavour to introduce rules and regulations which will apply to their particular advantage . . . There are never more than 6 to 10 per cent of the members in the running, and these few are usually at the top every year.
>
> (Shanks, 1926, p. 11)

Increasingly, however, the commodification and modernization of the industry led to the partial displacement of status and prestige by the rational accumulation of economic capital as its organizational principle. That is not to say that the industry was suddenly egalitarianized to the extent that small farmers could compete fairly with large landowning farmers. On the contrary, the large capitalist dairy businesses of the early twentieth century emerged disproportionately from the landowning gentry who were in a position to make the investments required to dominate the growing market. The rules of the game were changing, however, gradually but decisively, and capital increasingly became an end in itself rather than a means to prestige and status; the corollary was that the rationality of dairy production became increasingly utilitarian and instrumental, and this was reflected in the move towards breeding for increased yields. Thus the axis of competition in dairy breeding turned from a focus upon breeding 'trophy' animals for display, whose owners' superiority was attested to by the physical appearance of their pedigree animals, towards breeding highly productive animals capable of providing impressive milk yields. The conspicuous spectacle was displaced by the quiet rationality of statistical measurement.

There was a further tension in the use of milk records between the more long-term productivist principle of maximizing output and the established custom of selling cows when they were at their maximum market value.[56] Cows with good milk records attracted a higher price, whether as breeding animals or milk-producers, since this was seen as a guarantee of the future surplus value to be obtained either from themselves or their heifers – it was a measure of their as yet unrealized, future value. But even the carcasses of cows slaughtered for meat were also sold at higher prices when the farmer could produce milk records showing that it had been a prodigious milker. On the face of it this is a strange phenomenon, that the milk-producing history of a cow should impact in any way upon the value of its carcass. It suggests that the records came to be viewed as reliable statements of an animal's general quality, of its overall worth. Hence the statistical abstraction of milk yields were translated not just as intended into concrete use-values, from which further future values might be expected to emerge, but also directly into another abstraction, namely exchange-values, in a leap which eschewed the moment of production altogether. In this way the application of instrumentally rational formal abstraction was capable of leading to wholly irrational and unintended outcomes in practice. Indeed, the impact of official records upon the sale price of an animal was a major reason for the rapid growth in membership of official record societies during the years 1918–24. During this period it was not uncommon for non-pedigree cows with outstanding certified records to attract phenomenally high prices, which did much to recommend official recording to otherwise sceptical dairy farmers and those who were disinclined to recognize the financial benefits of increased productivity over the long term. This was short-lived, however, and as the effect of impressive milk records on prices stabilized, so the rate of increase in the membership of record societies slowed.[57]

The objective of increasing yields and that of maximizing short-term profits were incompatible, for where the custom of selling cows when they were at their most valuable was followed, it undermined the steady increases in yield which might otherwise result from a judicious use of milk records.[58] No doubt this tension broadly reflects the partial disarticulation of the moments of production and exchange within the valorization processes of capitalism, and the resultant duality of the animals themselves in their existence as commodities, being both means of production and exchange-values, both milk-producing machines and live 'stock'. That the productivist principle was increasingly paramount is evident in the gradually diminishing attention which was paid to the ultimate value of the carcass to the butcher at the end of an animal's productive life, a trend accelerated further by the emergence of specialist milk breeds, the value of which was determined not by their physical mass, but by the statistical profile contained in their milk records.

Surveillance and individuation

The practical impact of milk recording upon such key aspects of dairy management as breeding and feeding was ultimately dependent upon the epistemological shift which it brought to bear. The overriding object of recording was to enable the dairy farmer to know with some accuracy the yield of milk produced by the *individual* cows in his herd, and every practical use that could be made of milk records was built upon this knowledge. Previously, though a farmer would generally have had a rough idea of the *total* produce of his herd over a given period, with which he would have been acquainted by virtue of the regular sale of large batches to distributors, he would have had no way of knowing how this total was arrived at, in other words he would not have known in anything but the most impressionistic manner which of his cows were the most productive and which the least. Milk records made it possible for the first time for cows to be known formally and objectively as individual productive units. An integral part of this was that they were ascribed not only an abstract statistical identity, but an individual official history and a genealogy. Individual cows were able to be monitored both as producers with a definite history of productive output and as the descendants of animals which themselves had a recorded history of productivity; in this way their value as producers became a major determinant of how they were used. The practice of calf marking provides a concrete example of how this abstract individualization actually operated on the ground:

> In days gone by many farmers reared calves to replenish their herd, but took no steps to mark each calf, so that it could be identified later when it was old enough for service or came into the milking herd. This neglect is a great handicap to improvement by breeding, and one of the most helpful activities of Milk Recording Societies is the calf marking scheme, whereby calves are tattooed in the ear with a distinctive number within a few weeks of birth, and all particulars of breeding noted, so that each individual may be identified and its family history ascertained at any future date.
>
> (Macintosh, 1929, p. 67)

The marking of calves in this way was the material enactment of the form of individuation introduced by milk recording. Farmers were able to use the records both to identify inferior milkers and eliminate them, and to select the heaviest milkers whose progeny would be retained to replenish the herd.[59] This was distinct from the forms of individuation of animals which typically emerged from everyday farm work, and which tended to emerge from direct human–animal interaction, as well as from that involved in the breeding of show animals. Rather than the concrete individualization of direct interaction or the spectacular, qualitative differentiation of pedigree breeding, this was an abstract, quantitative differentiation, based narrowly upon the logic of production as mediated by milk recording. This individuation was not therefore a recognition of animal being as such, but rather was a negation, the reduction of animal being to a utilitarian abstraction. Cows became factors to be utilized in production, numerical units to be used in an equation, the solution of which was the maximization of milk yield.[60]

In addition to its transformative effects upon the selection, feeding, breeding and sale of stock, the work of milk recording exerted a pervasive influence upon farm management in general. A high degree of measurability was suddenly introduced into every aspect of dairy practice, the effects of which upon productive output could now be observed in an objective, external form. The slightest variation in practice could have significant ramifications for productivity and this compelled an attention to detail which was previously unknown in farm work.[61] Records of calving and service dates had to be kept with clockwork consistency; dry periods had to be properly regulated, food had to be given and milking undertaken at the exact intervals, and every detail carried out with meticulous precision:

> Milk Record Societies have also an incalculable influence on the dairy farmer himself. The work is of the kind which helps him to help himself, provides material for self-education . . . the cowman, also, is affected by a subtle influence; he is provided with an added interest in his work. He has on weighing days an easy means of measuring progress which he did not have before recording was introduced; and he is, almost unknown to himself, encouraged to milk more quickly, strip more thoroughly, feed more carefully, and treat his cows more quietly at all times in order to get better results.
>
> (Macintosh, 1929, p. 41)

Thus milk records introduced an insidious form of surveillance into farm work, in which the farmer's activity was subjected at all times to the requirements of increased productivity. The dairy farmer found himself perpetually in the gaze not of a human overseer but of an omniscient abstraction, the very principle of ceaselessly increasing productivity; hence he was obliged to self-regulate under the pressure of measurable 'progress'. In this way, milk recording constituted a political technology enabling disciplinary logic to enter into and shape the actions of the farmer at all times. Recording, in this sense, was a technique of panoptical surveillance and disciplinary power, operating by means of statistical knowledge rather than a configuration of material space, and with the singular aim of increasing productivity. It was closely bound up with, though not wholly reducible to, a transfer of power from farmers to the officials who certified the records, to the overseers of an administered society; it was therefore a crucial moment in the bureaucratization of dairy farming and in the antagonism between working farmers and the various officials and 'experts' who increasingly surveilled and micromanaged the industry. At one level this was a politico-epistemic and 'social' (human) transformation, but it also marked a reconfiguration of human–animal relations and a political operationalization of humanist discourse.

As a technology of individuation, milk recording paralleled developments which were occurring simultaneously in cowshed design, with individual cows being isolated as productive units, in the one case by material structures and in the other by statistical knowledge. This movement of disciplinary individuation followed the earlier movement of political aggregation which had followed the inception of agricultural statistics during the 1866 cattle plague; it did not mark a reversal of this earlier movement but a deepening and refinement of statistical knowledge and a shift in the attendant forms of politicization. Thus, in the first instance, statistics had led to the de-individualization of cattle, especially in relation to disease, allowing the aggregation of herds into a national cattle 'population' linked to the state and the national 'economy' as a territorial and economic unit. This 'population' could then be regulated and controlled as an objective unit and an element of national agricultural production. With the development of milk yield recording, herds were not merely abstracted 'upwards' into the national cattle 'population' as a strategic productive unit, but abstracted 'downwards' from the practical group level of the herd, which had for centuries been the 'natural' unit of dairy farming, to the level of individual animal bodies, which could then be regulated as productive units. This double regulatory movement, upwards to the level of a national economy and downwards to the level of individual bodies, is typical of the disciplinary power characteristic of modern governmentality, which seeks to act in detail upon all points at once and none in particular, on the liberal constitutional model of the rule of law.[62]

By entering into the farmer's activity, humanism was able to act upon the interface between species and to shape it materially through the mediation of such practices as breeding and feeding. The growing use of milk records in selective breeding, for example, was not merely a process of rationalization operating within a discrete sphere of 'economy', but was also a critical moment in the technical struggle of humanist discourse against the ontological problem of contingency in breeding. Modernity with its foundational dualism of 'nature' and 'culture' required that 'culture' defined as meaningful human action, including social production, be distinct from an orderly 'nature' both separate from and incapable of meaningfully impacting upon 'culture'. Just as the Renaissance Catholic Church had seized upon Descartes' vision of an orderly and rational universe comprehensible by the human intelligence as evidence of a divine will, thereby cementing its own authority, so modernity now seized upon milk recording as a means to render inheritance orderly, predicable and amenable to control, stripping the animal bodies which were the bearers of inherited traits of the power to undermine human reason, and thereby securing the modern sphere of 'culture'. Thus every technical advance of rationalization in dairy farming was simultaneously a theological advance for humanist discourse.

Conclusion: the rule of 'culture'

The second and third decades of the twentieth century witnessed a significant transformation of the British dairy industry, as milk production was subjected to a steadily deepening process of rationalization. This was heralded by the imposition in 1901 of a fixed legal standard for milk composition; though it affected production indirectly, this principally targeted the sphere of consumption by attempting to regulate milk at the point of sale. There was an irresolvable contradiction, however, between the 'pure' milk defined by this official standard and actual milk, that is to say, between the convenient legal fiction according to which all milk falling below a certain standard had to be adulterated, and a much more ambiguous material reality, in which the constituents of unadulterated milk were highly unpredictable. This contradiction meant that the milk standard was extremely difficult to implement, and the effort to enforce it effectively led to the potential criminalization of a significant proportion of dairy farmers. The result was to radically intensify the existing antagonism between farmers and the local officials and food inspectors who were increasingly entrusted with the regulation of the industry.

It was against this conflictual background that the National Milk Recording Scheme was put into operation in 1914, providing funding to cover the costs of setting up and administering milk recording societies. As a condition of this funding, it stipulated an exact method according to which all milk recording certified under the national scheme was to be carried out, and insisted that all annual figures be supplied to the Ministry of Agriculture. Unsurprisingly there was a significant degree of resistance to this extension of official regulation, though it was largely a passive and stoical resistance in the form of a wide-scale refusal to enter the scheme, and a correspondingly rapid growth in the keeping of private uncertified records. This was particularly prominent in the West Counties where dairy farming was a long-established way of life rather than simply a capitalist business. A countervailing factor, however, which worked in favour of a wider uptake of milk recording, was that the 1914 Scheme was a driving force for the massification of the industry. Indeed, the bureaucratization of dairy farming which official recording did so much to advance made its gains at the expense of the older form of organization, which was centred upon a rivalry for cups and trophies within an honours system dominated by the gentry and by the 'show' system of competitive breeding.

To some extent, therefore, the new system of national grants and government incentives raised the prospect of improving the technical development of dairy production across the board, to the benefit of smaller farmers who could not afford the luxury of breeding for competition and who would not have had any prospect of winning one of the yield contests that had hitherto provided the only system of incentives. For a few years this egalitarian dimension of the National Scheme seemed to be underlined by the extraordinarily high prices which could often be obtained for even non-pedigree cows with outstanding milk records. Though this was short-lived, it did much to speed the early uptake of recording. The short-term effect upon prices notwithstanding, in practice, milk recording did not have a particularly dramatic levelling effect, as it was disproportionately the wealthy who were in a position to dominate the expanding market, but it did lead to a shift in the stratifying principle of the industry away from status and prestige gained through competitive breeding and towards capital accumulation gained by steadily increasing productivity.

These relational and organizational transformations were underpinned by changes in dairy practice, which were in turn bound up with the political and epistemological shifts engendered by milk recording. At the most practical level, milk records provided farmers with a strictly quantitative and rational basis for decisions on feeding and breeding practices, allowing them both to manipulate rations in order to stimulate maximum milk production and to select for breeding only the most prodigious milkers. The ramifications of these changes were far-reaching indeed and went well beyond their immediate practical applications. They led to a gradual but decisive transformation of the relationship between farm practice and scientific dairy research, with farmers becoming increasingly dependent upon 'expert' nutritional research in their selection of foodstuffs; one result of this was the growing popularity of pre-mixed commercial concentrates. Milk records had an even more significant impact upon breeding practices, providing a vital new tool for farmers to make use of in their attempts to manipulate the dynamics of genetic inheritance.

At first glance there would appear to be a parallel between the epistemological impact of the microscope and scientific knowledge of microbes, as discussed in Chapters 3 and 4, and that of milk recording, insofar as both technologies rendered knowable a formerly invisible natural 'agency'. This is ultimately deceptive, however, for milk records did not reveal the agency of genes in the same way that the microscope had revealed the agency of microbes; in fact the agency of genes was still not understood as such – the gene itself had not been identified – and so breeders could not determine with any reliability what the outcome of their attempts at selection would be in any specific case; there was still a huge element of contingency involved. Thus farmers were like the hygienists before the discovery of the microbe; they had to try to act upon all points at once, to take account of the whole multiplicity of possible factors affecting inheritance, for they did not yet grasp the actual mechanisms of inheritance in their essential dynamics. Milk recording provided a technical means to overcome this dilemma: As with all statistical knowledge, it enabled the analysis of empirical regularities of outcome to substitute for an understanding of the motive forces underlying those outcomes, so that the detailed analysis of surface movements or external relations obviates the need for a depth model or theory of internal relations.

This is the nature of the 'science' of statistics, which by deducing patterns in apparently chaotic phenomena facilitates their instrumental control even in the absence of their understanding. Milk recording did not solve the problem of inheritance then, but it did represent a new way of knowing animals, allowing cows to be measured both as individual productive units and as sets of traits which might be passed on to their progeny. It enabled the bodies of animals to be known in a radically new way and to be measured and regulated across time, both the linear time of productivity and the cyclical time of generational reproduction. The development of artificial insemination decades later was the next step in this development and took it much further, as the species itself became an increasingly 'cultural' or 'economic' product into which the requirements of production were engineered at the level of corporeality. Each of these different ways of knowing were at the same time different techniques of power, or political technologies: Statistics are essentially panoptical and allied to disciplinary power and modern governmentality; they survey a complex multiplicity of outcomes and translate them by means of technical reduction into a numerical relationship, that is into a qualitatively homogeneous but quantitatively comparable form, graspable by a single observer and readily manipulable; or in other words into empirical 'data'. The form of power corresponding to this mode of knowledge requires an elaborate apparatus for its operation, since it must be capable of acting upon a wide range of 'variables', or practices. Depth-models, in contrast, which are characteristic of the natural sciences in general, but of physics above all, attempt by grasping the innermost dynamics or internal relations of a phenomena to influence a multiplicity of outcomes through action upon a single point.[63]

What is interesting is that these two formations of power-knowledge, apparently antithetical, are actually complementary; they are different aspects of a single regime of truth: disciplinary power, which is dispersed, omniscient and omnipresent, is the condition of possibility for the form of power entwined with depth-models, which is concentrated upon a single point of application. The latter requires that its object be precisely defined, the outcomes identified, the field of action delineated, and everything extraneous excluded, all of which is accomplished by disciplinary power. It seems possible to say, therefore, that disciplinary knowing and scientific knowing, the knowledge of the modern state and the knowledge of the laboratory, are inextricably intertwined; they are like negatives of each other. Modern governmentality defines a circumscribed sphere of action where politics is present, where power is legitimate, that is the realm of 'politics', while the political field itself is far wider, deeper, and subtler – invested within knowledges, material processes, disciplines and technologies as much as in classes, parties and parliaments. Science, on the other hand, defines a precise sphere of action where politics is absent, where knowledge is pure, that is, the realm of 'truth'; while the actual operation of power not only encircles this sphere but is intrinsic to it. The interpenetration of knowledge and power cuts both ways.

Where does humanism come into this? Humanist discourse defines the separation of nature from culture that underpins the supposed autonomy of science from politics and vice versa. Thus nature/culture–science/politics dualism is predicated upon the work of humanism in purifying the human/animal divide, the most marginal and problematic case for sustaining these other dualisms constitutive of modernity, its 'culture' and its 'politics'. Only if a domain of 'nature' exists as separate from an exclusively human domain of 'culture' or 'the social' can science and politics – and thus truth and power – remain distinct; they may be related *afterwards*, but the point is that they are conceived as distinct *in essence*, their relations are seen as external rather than internal. The condition of possibility of this separation of science and politics, of truth and power, is a separation of human and animal which at the deepest level is theological; the epistemological organization or 'constitution' of modernity is predicated upon a historically specific regime of species relations. It follows that human–nonhuman relations are in no way reducible to a relationship between 'culture' and 'nature'. In the case in question, for example, human 'social relations' were already deeply embedded within the form of life of the domesticated animals themselves, and becoming more so through the mediation of milk recording as a technology of rational selection, while equally human 'culture' was constituted through what was a historically specific regime of relations with nonhumans; in short, nature was shot through with culture and vice versa.

It is not enough to argue that ontological distinctions between nature and culture are misleading, however, one must also recognize that they are historically *necessary*; they both constitute and legitimize the regime of human–nonhuman relations which is the condition of possibility of modern 'culture', that is to say, of the way of knowing oneself as human which is at the heart of the modern condition. From this point of view, the natural and the social are not exclusive domains, since they have no essence which stop them merging into each other ceaselessly and thus losing their identity as discrete categories. Human–animal relations then are *species* relations, which are simultaneously 'social' and 'natural', and therefore neither. We enter into relations with animal species as social beings, but this sociality is a function of our natural form of life; or to put it another way, we enter into relations with animals as an animal species ourselves, but our existence *as* animals is deeply social. At the same time animal species enter into relations with humans *as* animal species, but as animal species which are no less social. Thus animal being *itself* includes specific forms of sociality, and far from the domain of the social being measured by the retreat of nature, sociality is always already natural; it is a configuration of nature, not a departure from it.

6 Beyond 'culture' and 'nature'

Towards a post-humanist knowledge

> Animals have been ominously ambiguous in their place in the doctrine of the autonomy of the human and natural sciences.
>
> (Haraway, 1991, p. 8)

> The future lies in human modesty about being human.
>
> (Miller, 2005, p. 36)

The preceding chapters have drawn upon the analytics of genealogy and generalized symmetry as critical strategies for decentring the modern 'social' and the 'cultural' subject, thereby making a more heterogeneous account of modernity possible, and one which creates the space for a non-anthropocentric analysis of human–nonhuman continuity and nature–culture interconstitution. Pursuing this agenda through a historical case study of the development of the British dairy industry, I have made a series of interconnected arguments and conclusions, both substantive and philosophical. I have argued that the category of 'culture' or 'the social', so pivotal in the organization of modernity as a form of order, far from being the ontological condition of human existence, has been historically constituted through the discourse of humanism and the material, epistemic and micro-political practices in which its categories are actualized. I have used this term 'humanism' to designate the discursive formation which holds that human beings inhabit an ontologically unique domain of intersubjective meaning and agency, incommensurable with the nonhuman world. This domain, I have suggested, is what the knowledge-practices of modernity, including those of social science, know as 'culture'. At the core of the foregoing analyses, therefore, has been the attempt to show that what we think of as 'culture' or 'the social' is enacted or performed into being within networks of historical practices which inscribe what it is to be human through a systematic othering of the nonhuman.

I have argued that our relations with nonhuman animals constitute an absent and unstable centre of this process, principally because animals are nonhumans of a peculiarly problematic kind for humanist and dualist ontology. Without recourse to Cartesian notions of mind, self-awareness or even sentience as such, I have argued that animals as bodily beings acting, behaving and existing in particular and sometimes unpredictable ways invite comparison with ourselves which are destabilizing, in a way that technological nonhumans do not. The closest thing to an exception would be computers as popular metaphors for the human 'mind'; but the difference is precisely that they *are* metaphors. This is not based upon any *a priori* humanism, but merely the practical observation that a machine and a brain are very different. For all the claims of artificial intelligence, who would seriously linger over the question of whether the metaphorical 'mind' of an abacus, however complex, is more like the human 'mind' than the actual brain of a primate is like the human brain; we are back to the absurd problem of satellites or gorillas. Perhaps this is a doctrinaire assertion of vitalism, and if so, then so be it. For only the hubris of an anthropocentric modernity could enable us to entertain the god-like notion that we more resemble our own technological creations than closely related animal species. It is merely another version of the theological principle that underlies the conceptual architecture of 'culture', namely the notion that we are self-creating beings, created in our own image.

The actual operation of the processes of purification which inscribe 'culture' and 'the social' upon the world has been the main object of my historical analysis. In order to trace these processes, I have focussed upon a specific field of practice, namely the networks of practices comprising the development of the modern British milk industry. I have suggested that these practices are illustrative of a wider but not identical or unilinear process in modernity, in which the division between human and nonhuman and its mapping onto 'culture' and 'nature' respectively were firmed up, institutionalized, tied into 'objective' processes, rendered less permeable, less negotiable, and made increasingly to appear as a fundamental, external and given reality. In a sense then I have described a process of reification; a process of systematic effacement of the ontological labour of purification which is the condition of possibility of 'culture'. This reification process is deeply political, but its politics is ontological; to render this ontological politics open to view and therefore contestable has been my fundamental objective. To this end, I have analysed the period from 1865 to around 1940 in dairy agriculture which, I have argued, saw an unprecedented realization and concretization of humanist categories through an emergent network of ontological-political technologies. Using historical documentary analysis I have traced the development of several of these technologies, showing how the progressive but uneven realization of humanist discourse was intimately connected to the struggle against diseases of cattle, to the expansion of the railway milk trade and the associated drive for clean milk in the towns and cities, and crucially to the battle against zoonotic tuberculosis transmissible through dairy milk. I have also argued that this transformation of human–nonhuman relations was closely bound up with the rationalization processes of modernity. In this sense, my analysis has been both a contribution to the theorization of modernity and an attempt to shift the terms of debate on modernity substantially towards the ontological politics of defining, classifying, and interpellating the 'human' vis-à-vis the living nonhuman other.

The fact that this critique has been significantly shaped by the insights of genealogical analysis and actor network theory is not without implications. Without doubt, genealogy is a powerful critical strategy for decentring the humanist subject–agent. Its consistently epistemological approach to history eschews the adoption of ontological explanation by deferring the moment of ontology indefinitely and focussing stubbornly upon the politics of knowledge. What I want to suggest, however, is that this in itself, though wholly necessary, is not in the end sufficient; it only clears the ground for a new conception of the human being – it is deconstructive but not reconstructive. This is problematic both theoretically and politically. First, the commitment to abstain from ontological discourse indefinitely can itself be grasped as an ontological position – it entails a certain vision both of what the world is like and of how we may claim to know it. In this sense, ontology is always there in the background, and the refusal to address the questions it poses will not lead to its disappearance. Hence, if ontology is always already epistemological, then the reverse is equally the case; epistemology is always inextricable from questions of ontology. Given this, it is clear that an insistence upon the endless subsumption of the ontological moment finally amounts to a reaffirmation of the essential logic of constructivism, that is, of the essentially 'cultural' nature of reality. But, as I have argued, constructivism is inescapably anthropocentric and is an extension of dualist ontology rather than a departure from it. To stop at decentring the modern humanist subject through historical epistemology then is not satisfactory. It seems to me to represent the best available strategy of auto-critique, enabling 'social' thought to think against itself, against its humanist assumptions, but in the end it leads inexorably to the ontological question of how we may understand the human differently, and in concrete terms; to hope to avoid ever addressing this question is futile.

Second, in the present climate, I believe it is intellectually irresponsible to contribute to the further dissolution of the humanist subject–agent without suggesting what may take its place, or at least pointing in the direction of a certain kind of 'post-human' future. There have always been powerful anti-humanist currents of various kinds embedded in modernity, and there are several reductive and tendentially authoritarian anti-humanisms, the resurgence or ascendancy of which it is not difficult to imagine, and into which it would be catastrophic to descend, however elliptically – not least eugenics and other forms of scientific racism, reductive biologism, genetic determinism, experimental behaviourism, game theory and rational choice theory, naming only those that spring instantly to mind. All involve a mechanistic conception of the human being. It is therefore crucial that the 'social' is not decentred only to be replaced by a vision of organisms as passive bearers of the agency of genes, or by the mechanistically rational and egoistic individual of 'economic man', both of which very much lie in wait for any final interdisciplinary collapse of 'social man'. Genealogical deconstruction therefore is not enough; it disarms humanism, but it cannot arm us adequately against some of its more dangerous alternatives.

Similarly, actor network theory, though unparalleled as a theoretical means of cutting through anthropocentrism, seems at times to be mired in the formalism of an underlying semiotics, underpinned by a distant but significant debt to Saussurean linguistics. Saussure of course posited that meaning emerged from a synchronic system of relations between signs, which was a relational theory of meaning.[1] But because 'signs' were imagined and constituted as objects of analysis by being abstracted from their material embeddedness in situations of practice, they are immaterial by nature; hence their relations are ideal, they are abstract 'mental pictures', and can therefore be theorized in formal terms.[2] In a recent article, John Law, who has been a key architect of ANT, describes it as a 'material semiotics', or in other words an approach which extends the insights of semiotics to the analysis of material entities.[3] In a parallel to Saussure, he argues that material orders emerge from the relations between materialities. Now I am all for rethinking materiality, but I don't believe that it can be accomplished by the use of a semiotic model. Materiality, after all, is neither ideal nor abstract; it is irreducibly concrete and particular, and for that reason I am not convinced that material relations are amenable to the formalism entailed by a semiotic approach. Indeed, the admirable ANT notion that society is 'materially heterogeneous' is finally incompatible with a 'material semiotic' approach, since semiotic relations are *not* in the end heterogeneous, by virtue of being formal abstractions. Material relations are neurophysiological, hydraulic, and organic; they are chemical, explosive, and tensile; they are genetic, electrical, and magnetic; they are corrosive, muscular, and visual; they are gravitational, auditory, and contagious. As Tim Mitchell has shown, these specific materialities make a difference to the phenomena we are accustomed to call 'social'; indeed, they constitute it.[4] What is more, they do not have to become signs and enter into a relational realm of 'meaning' in order to do so.[5] Social science therefore needs to develop not merely a vocabulary but an epistemology which recognizes the significance of the qualities of the material, and of the specificity and heterogeneity of materiality, rather than attempting once again to render materiality in terms of a semiotic conception of culture.

This residually semiotic approach to materiality is what has enabled ANT to subsume the specificity of animals under a conception of the nonhuman shaped by a preoccupation with technological objects. But this is anthropocentrism by other means; for while technologies can be thought of as human constructs, as artefacts into which human 'agency' has been engineered, even if it then becomes semi-autonomous, animals are obviously not in any similar sense human constructs,[6] and their 'agency' is therefore of a fundamentally more problematic kind for humanist ontology. If technologies are *human nonhumans*, then animals are *nonhuman nonhumans*. There is therefore a need for an account which views animals not only as irreducible to technocentric conceptions of the nonhuman, but as *the* critical case for any symmetrical account of the organization of modernity. I have tried to provide such an account here, by showing that the myriad techniques through which animals are ontologically managed or governed are crucial but often invisible elements in the constitution of the modern human subject as a 'cultural' being. What is needed then is not 'material semiotics', but something which rather than thinking materiality relationally, attempts to think it materially.[7] By grasping the material in terms of relational-semiotic logic, ANT tends inevitably towards a covert extension of constructivism; it does not really grasp materiality materialistically, and so is still modern, humanist, or asymmetrical in this respect.

For these reasons, my critique, though hitherto agnostic by necessity, cannot be agnostic to the very end; it must eventually address the question which arises, namely the question of how we may envisage a post-humanist turn, and indeed a post-humanist future. I would contend that wider currents in late modernity have created a historical conjuncture in which it is increasingly problematic to think 'the human' in terms of culture–nature dualism, whether in a constructivist or a naturalistic form. Historical conditions have precipitated a crisis of humanist modernity through the proliferation of agential natures: one thinks of climate change; of new forms of zoonotic disease and antibiotic resistance; of the emergence of new social movements around various 'politics of nature'; of the growth in public concern for animal welfare in food production networks; of the social and political implications of genomics. In short, what Latour calls 'the proliferation of hybrids' is now overwhelming the work of purification which was always necessary to sustain the core asymmetry of modern dualism. If that were not so, then this book would not have been conceivable. In this context, it is increasingly inadequate for social science to continue in its dualist epistemology, its religious attachment to the fallacy of the uniquely human subject–agent, hence to the immanence of 'culture' or the 'social', which cannot stand up in the face of an increasingly pervasive and encroaching nature. The bastion of the 'social' will not survive while remaining the opposite of the natural; it must enter into nature, as it were 'dialectically', and complexify it beyond recognition so that what remains is neither 'nature' nor 'society' as modernity has known them.

To an extent this has been the work of science and technology studies or STS, but ultimately this tradition too remains asymmetrical, because in practice it has often been reluctant to allow the mutual interconstitution of nature and culture to cut as deeply in both directions. For example, Latour's recent notion of 'association' as an alternative to the 'social', which better captures the human–nonhuman heterogeneity of life, is undoubtedly a step forward, particularly in recognizing that the 'social' has become so replete with humanist assumptions that new terminology is now perhaps called for in order to think 'sociality' in the radically different way required. 'Association', however, like the 'network' and 'assemblage' which preceded it, is still semiotic and even structuralist in its essential logic, for it instinctively identifies the *relations between* entities as the defining moment of 'societies–natures', which give shape and definition to the entities themselves. This is the basic strategic assumption of semiotic anti-essentialism, but it is also the critical error of formalist thinking. For in focussing upon the relations *between* entities as though *they* are what really matter, one both inscribes an untenable prior separation between entities and their inter-relations, and neglects the significance of the material properties of the entities themselves – what might once have been called their 'content' or being – and therefore the material specificity of their relations, treating them instead as 'empty' and interchangeable points on an abstract mental map, hence as though homogeneous.

This reduction to abstract or formal relationality is made especially visible by an analysis of human–animal relations, and made visible as an underlying anthropocentrism; the reason being that these relations are utterly mediated by the particular corporeality of the species in question, which is in no primary sense a product of the relations themselves. What matters, for example, in shaping the relationship between human beings and cows, is not only a long history of domestication and breeding and human agricultural practices, but also certain basic elements of 'cowness' and 'humanness' which have persisted, and which have been the conditions of possibility of everything else. The modern cow is without doubt a relational construct, in innumerable ways, but if it were not also a mammal which turns grass into milk, then none of this would matter. Thus there is something in 'cowness' which is the emergent product of an evolutionary history which, though it can itself be understood as relational rather than static, involves relations operative over a time scale which dwarfs that of the socio-technical. It is not that these modes of relationality are different in kind, or in absolute terms, but rather, as Tim Newton has suggested, that their temporal dynamics are so far apart as to be very nearly incommensurable.[8] As a result, when an organism confronts a socio-technical assemblage, it does so by and large as something relatively fixed, immobile, and for all practical purposes given in nature; hence the two cannot without reduction be theoretically dissolved into one overarching relational ontology. It is in this respect that nonhuman animals are peculiarly resistant to any ultimate subsumption under the logic of semiotics. I would therefore suggest that they demand to be grasped not in terms of a poststructuralist material semiotics, to which they are finally irreducible, but in terms of some more concrete notion of *species practice*, something like a materialist conception of a complex *form of life*. This would have the distinct advantage of being rooted in actual activity mediated by specific corporeality-in-time, rather than in an abstract relationality.

The notion of a 'form of life' derives from the later philosophy of Wittgenstein, for whom it was accompanied by what is usually taken to be a fairly emphatic anthropocentrism, summed up in his well-known argument that 'If a lion could talk, we could not understand him.'[9] For this reason it may not seem a promising concept with which to begin to develop a way to think inter-species relations non-anthropocentrically. But Wittgenstein also maintained that linguistic utterances became meaningful in the context of an ensemble of lived activities, of observable or 'public' behaviours; their meaning did not depend upon the 'private' mental states of the speakers, nor upon the functioning of words as 'signs' in some system.[10] On these grounds he was able to show that there could be no such thing as a 'private language'. Language, for Wittgenstein, was not to be understood as a system of representation, but as an aspect of activity or behaviour, as a form of life. This is crucial because it provides the basis of both a critique and an alternative to relational-semiotic theories of meaning, with their underlying abstraction and formalism, positing meaning instead as something emergent from and embedded within practices, and therefore as constitutively material, embodied, and able to be inferred from observation.

Extending this logic, one can begin to imagine how forms of life might overlap and interlock within and across various practices in ways which create the possibility of degrees of commensurability, that is, of meaningful relations across species boundaries, precisely because these relations emerge from embodied activities and ways of being and doing things, rather than from some synchronic, entirely immanent, and exclusively human system of linguistic signs. We share with nonhuman animals the fact of being corporeal beings-in-the-world, and though our embodiments differ significantly, this only mediates and complexifies our inter-species relations and does not render them meaningless or incommensurable. What emerges is a much more complex picture of *degrees* of commensurability, modes of relationality and association with their own forms of meaningfulness, mediated by concrete species corporealities, rather than anything as simple and binary as a culture/nature, human/nonhuman, or human/animal distinction would suggest. How then might the post-human be imagined in these terms? The starting point would necessarily be rather different from that of ANT. Instead of dissolving the human into networks of relations between ostensibly heterogeneous but semiotically modelled (and therefore substantively homogeneous) entities, a sociology of species practice would dissolve the cultural subject–agent of humanist discourse into actual living humans.

From this point of view, it becomes essential to acknowledge the overwhelming significance of the fact that humans are animals, nothing more; but for that matter, as Nietzsche realized so profoundly, nothing less.[11] Sociology risks irrelevance if it continues to define its object of knowledge in terms of a human/animal distinction, however well obscured and deeply buried within its conceptual architecture; what is needed is a sociology of the *human animal*. This would not, however, be a reductive 'biological determinism' because, according to the logic of generalized symmetry, it would have to be accompanied by the equal recognition that animals are profoundly social, and that what we anthropocentrically think of as 'culture' is an intrinsic part of species life, both species-specific and bound to natural premises. Thus, both dimensions of this deconstruction must be thought simultaneously, in something like a 'negative dialectic',[12] if one is to avoid the trap of a positive reduction to one or other pole of the humanist dichotomy. It follows that 'culture' is not uniquely human, nor is it a departure from a deterministic 'nature', but rather culture is relative to forms of life or modes of existence, which are not 'natural' as such.

It is no use building castles in the sky, however. So the question must be, in what form of actually existing knowledge-practices can the seeds of a conception of species practice be detected? And what I want to suggest, in fact, is that something approaching this can be glimpsed embryonically in the emerging fields of *social* and *cultural* ethology and primatology. But as self-identified 'natural' sciences, these disciplines can still be philosophically naive in their epistemologies and their conceptions of their own knowledge-practices. For this reason I envisage a post-humanist turn in terms of a strategic rapprochement between post-positivist social science and social-cultural ethology, an animal turn then, even an *ethological turn*, but one which is undertaken in the spirit of an anti-reductive epistemology, a symmetrical and dialectical animal sociology. In any case, the explanatory simplicities of our notions of the 'natural' and the 'social' will have to give way to a recognition of species complexity if we are to really grasp our world and ourselves in a way that dispenses with the last vestiges of a theological conception of the human being.

Such a vision is still controversial, so I should say more about what I mean by it. I certainly do not imply that social scientific explanation should begin to adopt the kinds of biological reductivism developed in the past, and in some quarters still being developed, by the science of animal behaviour, and seek to apply it to humans. The ideological nature of much of this kind of work is powerfully critiqued by Donna Haraway, who has argued that 'One area of the bio-behavioural sciences has been unusually important in the construction of oppressive theories of the body political: animal sociology, or the science of animal groups.'[13] Tracing the historical trajectory of this 'animal sociology', she points out that:

> First, its subject and procedures developed so as to span the nature-culture split at precisely the time in American intellectual history, between 1920 and 1940, when the ideology of the autonomy of the social sciences had at last gained acceptance . . . Intrinsic to the new liberal relations of natural and social disciplines was the project of human engineering – that is, the project of design and management of human affairs for efficient, rational functioning in a scientifically ordered society. Animals played an important role in this project.
>
> (Haraway, 1991, p. 11)

> Second, animal sociology has been central in the development of the most thorough naturalization of the patriarchal division of authority in the body politic and in the reduction of the body politics to sexual physiology.
>
> (ibid., p. 12)

These arguments concerning the ideological form, content, and function of a great deal of the science of animal behaviour, historically speaking, are absolutely critical, and must be borne constantly in mind in any social scientific engagement with ethology. We must not be seduced by any fantasy of a return to natural-scientific objectivity as an easy answer to the epistemological dilemmas of post-positivist and post-humanist social science. However, such considerations should not be used to justify another retreat into culturalism, with animals treated as merely the passive objects of human social relations or 'culture'. This rather unimaginative default position still essentially underpins a good deal of the work presently being published under the rubric of the 'sociology of human–animal relations',[14] which effectively shies away from the unsettling dilemmas of how to understand our embodied, lived, and indeed existential relations with other animals, in favour of an approach which grasps these relations only by dissolving them into what is ultimately an anthropocentric abstraction.

With this in mind, while I respect the necessity of Haraway's argument, I want to emphasize that it was fundamentally the prior separation of 'culture' and 'nature' and the equation of the 'cultural' with the human subject, hence with agency and choice, and of the 'natural' with the nonhuman object, hence with the inevitable and determined, that *enabled* the science of animal behaviour to be used to 'naturalize' human social relations of domination, in the sense of making them appear given and 'objective'. Naturalization itself therefore is not the problem; it becomes a problem only in the context of a modern ontology which casts 'nature' as the other of human 'culture'. This I believe is still too often missed, or perhaps left deliberately unacknowledged by a social science interested in maintaining its own authority by reasserting the autonomy of culture,[15] namely that 'culture' and cultural knowledge are part of the same problem as 'nature' and natural knowledge; they are not the solution.

Haraway's hostility to the 'science of animal behaviour' is based upon her identification of a 'principle of domination' at the heart of this science, which she believes has shaped and conditioned its knowledge in the interests of human relations of domination and their legitimation. More recent developments in social ethology, however, suggest that this characterization, though it has some accuracy historically, is less and less adequate as a description of contemporary social ethology.[16] At the very least there are now significant currents of thought within the study of animal behaviour which cannot convincingly be characterized in these terms. Certainly the theme of domination is still present in some of this work, as indeed it must be; for any study of animal and particularly primate behaviour which attempted to elide the realities of endemic power struggles, widespread aggression and relationships of domination, would be a very selective one indeed. Increasingly, though, such elements are viewed not in terms of some overarching vision of a nature 'red in tooth and claw', which can be contrasted with human 'civilization' and 'culture' while at the same time being used to 'naturalize' relations of domination in human society, but as part of a much more complex set of relations and practices.

This turn to complexity has led to a growing awareness of the interfusion of co-operation and competition, of elaborate status hierarchies and rituals, sophisticated organization and communication, population-specific practices, geographically differentiated modes of subsistence and interaction with the environment, the significance of non-functional behaviour, and evidence of elaborate symbol use and abstract reasoning.[17] In other words, animals are now beginning to be viewed by progressive ethologists as fully 'social', indeed as 'cultural' beings, rather than as instinct or gene-driven biological machines. Though this perspective has also made significant advances in studying non-primates including birds and dolphins, it is most advanced in cultural primatology:

> Cultural primatology is analogous to cultural anthropology, as a subset of investigations interested in the culture (as opposed to the anatomy, ecology, genetics, or physiology) of nonhuman primates. However, whereas cultural primatologists assume that nonhuman primates are cultural creatures, most cultural anthropologists presume that only humans have culture. Both cannot be right.
>
> (McGrew, 2003, p. 420)

This is a very significant and very promising development indeed, with huge implications for social science, not only because if it continues and comes to dominate the field, this new paradigm will effectively undermine what has been one of the most potent sources of reductive theories of *human* nature, but also because it represents an unparalleled opportunity to dissolve the foundational dualism of nature and culture which made such reductivism possible, in a way that radically transforms our understanding of both the world and ourselves. Far from the ambiguity of animals in the doctrine of the autonomy of the social and natural sciences being ominous then, as Haraway believes, it is the source of our best hope of change.

As social scientists we now *need* animals, and studies of animal 'social' behaviour, to help us to transform our ways of knowing, because our own foundational pact with modern dualism runs too deep. Our notions of the human being are too thoroughly bound up with a humanist ontology; thus to re-imagine our human selves we must re-imagine our most problematic others. At the same time, social ethologists increasingly need anthropologists, sociologists and linguists to aid them in developing new conceptual and epistemological frameworks, reflexive methodologies and practices for studying these nonhuman 'cultures'. An example of the kind of latent but fundamental problem in social ethology which social scientists might well be better equipped to navigate is the fact that, in seeking to determine the extent to which various nonhuman animals are cultural beings, social ethologists have typically sought to operationalize concepts of 'culture' developed in reference to humans. There is a certain tautology to this, for the use of such concepts can only logically tell ethologists the extent to which nonhumans are *like humans*, rather than the extent to which they are 'cultural' as such. The assumption, in other words, is still that humans are the benchmark of what it is to be a 'cultural' being. To return to Wittgenstein's phrase, we are not asking whether the lion can speak, but merely whether it speaks our language.

If we take the 'cultural' nature of nonhumans as a strategic starting point, however, on the grounds that, as I have argued at length in the foregoing chapters, 'culture' is always a matter of historical attribution rather than a 'real' exclusive club, the membership of which might be established empirically, then it is clear that such different beings will have very different ways of being cultural, rendering human measures of 'cultural-ness' redundant. If culture is anything, then it is both immanent and relative to forms of life. The problem then becomes analogous to issues around the politics of knowledge of subjugated cultures and the voice of the subaltern in post-colonial, feminist, and working-class theory. More importantly it points to the ultimate limitations of the strategy of extending the boundaries of 'culture' to nonhumans, given that the concept is humanist in its innermost logic: We must surpass the anthropocentrism of 'culture' altogether.

In a self-critical engagement with nonhuman animals and social ethology lies the key to our post-humanist future, for this alone will compel social science to reconcile the irreconcilable, to go beyond the dualism of 'culture' and 'nature', and to do so without recourse to a constructivist epistemology. Only then can we begin really to understand that like other animals we are at once products of evolution *and* association, we are both bearers of genetic traits *and* creative individuals, we are creatures of instinct *and* learning, we are mammals, primates, *and* we are communicative beings; most crucially though, we are all these things and more *at once, simultaneously, on one 'level'*, not on two. The complex species practices of animals help to show us that in the end it is not a question of the relations between nature and culture, but of their identity: there *is* no 'nature' and 'culture'; there is only life, and there are forms of life. There must be a corresponding politics.

To reach this point, however, as I have tried to show, it is necessary to go beyond the empirical or theoretical extension of human 'culture' into nonhuman 'nature' and vice versa. Such moves are tactically important in the struggle to destabilize humanist discourse, whether in the form of the supposed autonomy of natural or social science; but ultimately the project of showing that formerly 'natural' phenomena are also 'cultural' and that formerly 'cultural' phenomena are also 'natural', can only take us so far. We need to excavate the role that these categories and their corresponding practices have played in the ordering of modern ways of knowing and living 'humanity', and to recognize the ontological and existential politics of this organization of our form of life. Perhaps then, at last, we can begin to change it.

Notes

Preface

1 This phrase is taken from Bruno Latour's *Politics of Nature: How to Bring the Sciences into Democracy*, Cambridge, MA: Harvard University Press, 2004.

2 For an historical-cultural analysis of the role of meat production in modernity, see Paula Young Lee (ed.) *Meat, Modernity and the Rise of the Slaughterhouse*, Durham, NC: University of New Hampshire (Library Binding), 2008.

Introduction: modernity, humanity and nonhumans

1 For the elegant phrase 'human exceptionalism', I am indebted to Ben Campbell's paper 'How bio-technology makes human kinship with nature visible', presented at *Animals and Science: Anthropological Approaches*, University of Manchester, 24 June 2005.

2 Gilles Deleuze and Félix Guattari, *A Thousand Plateaus: Capitalism and Schizophrenia*, London: Continuum, 2004.

3 Michel Foucault, *The Order of Things: An Archaeology of the Human Sciences*, London: Tavistock, 1970, p. xxii.

4 This roughly corresponds to the transformation which John Pickstone has described in the following terms: 'It was chiefly from the later nineteenth century that (some) academics, (some) industrialists and (some) state agencies began to create the networks for the systematic production of novelty which were increasingly characteristic of the twentieth century', in *Ways of Knowing: A New History of Science, Technology and Medicine*, Manchester: Manchester University Press, 2000, p. 3.

5 Bruno Latour, *We Have Never Been Modern*, Harlow: Pearson, 1993, pp. 10–13.

6 In stressing the fundamental importance to modern thought and culture of the human/animal distinction, I am in agreement with Jacques Derrida, *The Animal That Therefore I Am*, New York: Fordham University Press, 2008. Unlike Derrida, however, I am not content to remain at the level of abstract discourse, but instead want to trace the processes of making this distinction as they have operated historically and in practice.

7 According to Foucault:

> One has to dispense with the constituent subject, to get rid of the subject itself, that's to say, to arrive at an analysis which can account for the constitution of the subject within a historical framework. And this is what I would call a genealogy, that is, a form of history which can account for the constitution of knowledges, discourses, domains of objects, etc., without having to make reference to a subject which is either transcendental in relation to

> the field of events or runs in its empty sameness throughout the course of history.
>
> (Foucault, 1991, p. 59)

1 The anthropocentrism of 'culture'

1 On this, see Andrew Abbott, *Chaos of Disciplines*, Chicago: Chicago University Press, 2001.
2 In fact, it was only with the publication of Anthony Giddens' *Capitalism and Modern Social Theory* in 1971 that Marx began to become widely regarded as a 'sociologist', a label he would without doubt have vigorously repudiated.
3 Terry Eagleton, *The Idea of Culture*, Oxford: Blackwell, 2000, p. 1.
4 *Chambers Dictionary of Etymology*, Edinburgh: Chambers, 1988, p. 241.
5 Eagleton, *The Idea of Culture*, p. 1.
6 *Chambers Dictionary of Etymology*, p. 1029.
7 Ibid.
8 Ibid.
9 This is very close to the synchronic-semiotic model of a language, a model which 'culture' has increasingly emulated in the wake of the linguistic turn.
10 Peter Brooker, *A Concise Glossary of Cultural Theory*, London: Arnold, 1999, p. 51.
11 Raymond Williams, *Culture*, London: Collins, 1981, p. 13.
12 This simultaneous rejection and reification of science and nature by constructivist social science are criticized by Donna Haraway in the following terms:

> We have allowed our distance from science and technology to lead us to misunderstand the status and function of natural knowledge. We have accepted at face value the traditional liberal ideology of social scientists in the twentieth century that maintains a deep and necessary split between nature and culture and between the forms of knowledge relating to those two putatively irreconcilable realms.
>
> (1991, p. 8)

13 See Andrew Abbott, *The System of Professions: An Essay on the Division of Expert Labour*, Chicago: University of Chicago Press, 1988.
14 Something which Ted Benton and Peter Dickens, among others, have consistently opposed. It is no surprise that both are indebted to Marx. For perhaps the most influential (or controversial) statement, see Ted Benton, 'Biology and Social Science: Why the Return of the Repressed Should Be Given a Cautious Welcome', *Sociology*, 1991, 25(1): 1–29; or more recently Sandra Moog and Rob Stones (eds) *Nature, Social Relations and Human Needs*, Houndmills: Palgrave-Macmillan, 2008; for Dickens, see *Society and Nature: Changing Our Environment, Changing Ourselves*, Cambridge: Polity, 2004.
15 For a classic defence of this and its deployment within the social sciences, see Roy Bhaskar, *The Possibility of Naturalism*, London: Routledge, 1988; and more recently, Bhaskar, *A Realist Theory of Science*, London: Verso, 2009.
16 See Bruno Latour, *We Have Never Been Modern*, Harlow: Pearson, 1993.
17 Nikolas Rose, *Powers of Freedom: Reframing Political Thought*, Cambridge: Cambridge University Press, 1999, p. 3.
18 One exception would be Donna Haraway, who in as early a work as *Primate Visions* recognized something close to this possibility. I discuss her work in the concluding chapter.
19 For more on this, see Alfred Schmidt, *The Concept of Nature in Marx*, London: New Left Books, 1971. For a classic statement of the historicist Marxist view of nature, see Georg Lukács:

> Nature is a societal category. That is to say, whatever is held to be natural at a given stage of social development, however this nature is related to man and whatever his involvement with it takes, i.e. nature's form, its content, its range and its objectivity are all socially conditioned.
>
> (1971, p. 234)

20 Schutz's hugely influential *The Phenomenology of the Social World*, Evanston, IL: Northwestern University Press, 1967, was critical in this.
21 See Robert Young, 'Nature', in Tom Bottomore (ed.) *Dictionary of Marxist Thought*, Oxford: Blackwell, 1991, p. 399.
22 Terrell Carver, 'Darwinism', in Bottomore (ed.) *Marxist Thought*, p. 131.
23 Francis Wheen, *Karl Marx*, London: Fourth Estate, 2000, p. 364.
24 Ibid.
25 Terrell Carver, 'Darwinism', in Bottomore (ed.) *Marxist Thought*, p. 131.
26 Ibid.
27 See Frans de Waal and Peter Tyack (eds) *Animal Social Complexity: Intelligence, Culture, and Individualised Societies*, Cambridge, MA: Harvard University Press, 2003.
28 That this is Marx's position is made very clear in Harry Braverman's influential reflections upon the Marxist conception of labour in his *Labour and Monopoly Capital*, New York: Monthly Review Press, 1974.
29 According to McGrew:

> All that we know of knowledge, meaning, and emotion is based on behaviour. We have no direct access to human or any other minds, so all is inference. Whether or not we are any better at divining human than nonhuman minds is debatable. We share the perceptual world of our fellow humans (advantage) but we are also susceptible to their mendacity (disadvantage).
>
> (2003, p. 421)

30 See the considerable literature in *Historical Materialism: Research in Critical Marxist Theory* on 'systematic dialectics' and 'dialectics of value', including notably: Tony Smith, 'The Relevance of Systematic Dialectics to Marxian Thought', 1999, no. 4, pp. 215–40; Tony Smith, 'A Reply to Fine, Lapavitsas and Milonakis', 2000, no. 6, pp. 139–44; Andrew Kilman, 'Marx's Concept of Intrinsic Value', 2000, no. 6, pp. 89–113; and Geert Reuten, 'The Interconnection of Systematic Dialectics and Historical Materialism', 2000, no. 7, pp. 137–65.
31 For more on this, see Jacques Derrida, *Spectres of Marx*, New York: Routledge, 1994.
32 See Robert Young, 'Nature', in Tom Bottomore (ed.) *Dictionary of Marxist Thought*, Oxford: Blackwell, 1991.
33 H. H. Gerth and C. Wright Mills, 'Intellectual Orientations', in H. H. Gerth and C. Wright Mills (eds) *From Max Weber: Essays in Sociology*, London: Routledge, 1991, p. 56.
34 See Andrew Pickering, *The Mangle of Practice: Time, Agency and Science*, Chicago: University of Chicago Press, 1995.
35 As a range of social thinkers on nature, from Peter Dickens to Noel Castree, have persistently argued. For Dickens, see *Society and Nature*, Cambridge: Polity, 2004; for Castree, see Noel Castree and Bruce Braun (eds) *Social Nature: Theory, Practice and Politics*, Oxford: Wiley-Blackwell, 2001.
36 Latour, *We Have Never Been Modern*.
37 The defining text of this approach is Steven Shapin and Simon Schaffer, *Leviathan and the Air Pump: Hobbes, Boyle, and the Experimental Life*, Princeton, NJ: Princeton University Press, 1985.

38 Christopher Pinney, 'Things Happen: Or, From Which Moment Does That Object Come?', in Daniel Miller (ed.) *Materiality*, Durham, NC: Duke University Press, 2005, p. 258.
39 This term is derived from Clifton Bryant (1979), who speaks of a 'zoological connection'.
40 See John Law, 'Introduction: Monsters, Machines and Sociotechnical Relations', in John Law (ed.) *A Sociology of Monsters? Essays on Power, Technology and Domination*, London: Routledge, 1991, pp. 1–23. Also see W. Bijker and John Law (eds) *Shaping Technology, Building Society: Studies in Sociotechnical Change*, Cambridge, MA: MIT Press, 1992,
41 Karl Marx 'Economic and Philosophical Manuscripts' and 'Capital', in David Mclellan (ed.) *Karl Marx: Selected Writings*, Oxford: Oxford University Press, 1977, pp, 83–5, 435–7.
42 I have in mind, for example, many of the contributors to the conference 'Materializing the Subject: Phenomenological and post-ANT Objects in the Social Sciences', at the Centre for Research on Socio-Cultural Change (CRESC), University of Manchester, 26–27 February 2009. The perceived need to go 'beyond' ANT could no doubt also be attributed, at least in part, to contemporary social science's need for an appearance of constant novelty, with which the better to sustain its infrastructural basis. Such a dynamic demands a rate of built-in-obsolescence for theories which has very little to do with the actual discrediting or surpassing of existing theories.
43 Daniel Miller, 'Materiality', in Miller (ed.) *Materiality*, pp. 36–7.
44 Voluminous work in animal studies has stressed the key role of nonhuman animals in human identity construction and cultural formation more widely. A very selective sample of representative texts would include Steve Baker, *Picturing the Beast: Animals, Identity and Representation*, Illinois: University of Illinois Press, 2001; Lorraine Daston and Gregg Mitman (eds) *Thinking with Animals: New Perspectives in Anthropomorphism*, New York: Columbia University Press, 2006; Donna Haraway, *The Companion Species Manifesto: Dogs, People, and Significant Otherness*, Chicago: University of Chicago Press, 2003.
45 Michel Callon, 'Some Elements of a Sociology of Translation: Domestication of the Scallops and the Fisherman of St. Brieuc Bay', in John Law (ed.) *Power, Action and Belief: A New Sociology of Knowledge*, Keele and London: *Sociological Review* and Routledge, 1986, pp. 196–233.
46 Malcolm Ashmore, 'Behaviour Modification of a Catflap: A Contribution to the Sociology of Things', *Kennis en Methode*, 17 (1993): 214–29.
47 John Law and Annemarie Mol, 'The Actor Enacted: Cumbrian Sheep in 2001', in Lambros Malafouris and Carl Knappett (eds) *Material Agency: Towards a Non-Anthropocentric Approach*, New York: Springer, 2007.
48 Mike Michael, *Constructing Identities: The Social, the Nonhuman, and Change*, London: Sage, 1996, p. 137.
49 Annemarie Mol, 'Ontological Politics: A Word and Some Questions', in John Law and John Hassard (eds) *Actor Network and After*, Oxford and Keele: Blackwell and *Sociological Review*, 1999, pp. 74–89.
50 Michel Foucault, *Discipline and Punish: The Birth of the Prison*, London: Penguin, 1991, pp. 142–3.
51 Michel Foucault, 'Governmentality', in Graham Burchell *et al.* (eds) *The Foucault Effect: Studies in Governmentality*, London: Harvester Wheatsheaf, 1991.
52 Michel Foucault, 'Truth and Power', in Paul Rabinow (ed.) *The Foucault Reader: An Introduction to Foucault's Thought*, London: Penguin, 1991, p. 60.
53 See, for example, the articles by John Law, Kristin Asdal and myself in *Distinktion: Scandinavian Journal of Social Theory*, 2008, No. 16, special issue on 'The Technologies of Politics'.

2 Milk and modernity Part I: commodities, networks and monopolies

1 David Grigg, *English Agriculture: An Historical Perspective*, Oxford: Blackwell, 1989, pp. 216, 219.
2 Alan Jenkins, *Drinka Pinta: The Story of Milk and the Industry that Serves It*, London: Heinemann, 1970, p. 50.
3 L. W. Moffitt, *England on the Eve of the Industrial Revolution*, 1925, pp. 69–70, 120. For an interesting study of regional variation in the rate of abolition of town cows, see H. Hill, 'Liverpool – Last Stronghold of Town Cowkeepers', *Dairy Engineering*, April 1956, pp. 107–10.
4 G. E. Fussell, *The English Dairy Farmer, 1500–1900*, London: Frank Cass & Co, 1966, p. 310.
5 According to Horner:

> It is well known that prior to the disastrous cattle plague years (1865–7) London was dependent upon the cows kept in the Metropolitan area, or within a radius of a very few miles thereof; that in consequence of the slaughter of so many of the animals it became necessary to seek fresh fields and pastures new.
>
> (1909, p. 43)

6 Grigg, *English Agriculture*, p. 218.
7 Jenkins, *Drinka Pinta*, p. 49.
8 See Karl Marx, *Capital: A Critique of Political Economy*, Volume 1, London: Penguin, 1976, p. 129.
9 These ideas are derived from those of Guy Debord, *The Society of the Spectacle*, London: Rebel Press/Dark Star, 1977, pp. 147, 149, 163.
10 'With the application of the refrigerator an impetus was given to what was practically a new industry.' Horner, 'The Milk Supply', p. 43.
11 This is not especially cold by today's standards, but the limitations of the early cooling technology and the rudimentary state of bacteriological knowledge meant that 40 degrees Fahrenheit was the generally accepted temperature then in use.
12 Fussell, *The English Dairy Farmer*, p. 311.
13 Ibid.
14 According to Sadler:

> At some loading stations churns of milk have to be carried across the line by the farmer or his men to save the company the expense of providing an approach on the loading side of the station. At others, churns of milk have to be carried along the platform hundreds of yards to get them to the van.
>
> (1907, p. 19)

15 Horner, 'The Milk Supply', p. 38.
16 One milk distributor assessed the matter as follows:

> The question of sealing or locking churns for transit by rail is a debatable subject, and if the principal reason for the adoption of it be the prevention of tampering with the milk during transit, then, except that it would make it impossible to dip into the milk a dirty or unsanitary vessel, it is, in my opinion, useless. While cases of tampering with milk in transit have been known, it is fair to say that such cases are rare, and, compared with the huge volume of traffic, almost infinitesimal. The railway servants as a body must not be condemned for the peculations of a few.
>
> (Horner, 1909, p. 37)

17 Hugh A. Macewen, *The Public Milk Supply*, London: Blackie, 1910, pp. 63–4.
18 According to Davies:

> For years there has been an agitation afloat to bring in one uniform system of weights and measures for the whole country. The enormous variety of weights and measures, sometimes even in the same county, is perplexing in the extreme, and must prove a source of loss to farmers, who seldom consult tables of equivalents, and who, therefore, can only with the greatest difficulty ascertain from market quotations what the current value of their produce is in other markets. It is generally understood that only imperial weights and measures are really legal, and that if we bought and sold strictly in accordance with the law, uniformity would at once be established. Still, everyone knows that local customs die hard, and it would seem impossible to persuade the users of these innumerable variety of weights and measures to discard their own local customs with which they have been reared and to adopt the standards of some other particular area of which they know little.
>
> (Davies, 1906, pp. 90–1)

19 Ibid., p. 91.
20 For an influential analysis of the politics of measurement, see Timothy Mitchell, *Colonising Egypt*, Berkeley, CA: University of California Press, 1992, especially Chapter 2 with its discussion of 'enframing', as well as his *Rule of Experts* (2002).
21 Davies, 'Legislation Affecting the British Farmer', p. 91.
22 Ibid.
23 Fussell, *The English Dairy Farmer*, p. 310.
24 Note that I do not posit an absolute natural limit here: all such absolutes tend to recede over time with sufficient technoscientific intervention, albeit in ever smaller increments and with ever greater associated risks. Cow biology is no exception. But the key point here is that in the early twentieth century twice-daily milking was *experienced* as an absolute given.
25 As Horner pointed out:

> It is very desirable that the train service should be better adapted to the needs of the [milk] traffic, which is constant, and not subject to the fluctuations of ordinary passenger and commercial traffic.
>
> (1909, p. 38)

26 Fussell, *The English Dairy Farmer*, p. 317.
27 Horner, 'The Milk Supply', p. 38.
28 Fussell, *The English Dairy Farmer*, p. 320.
29 Ibid., p. 314.
30 Ibid., p. 321.
31 Grigg, *English Agriculture*, p. 220.
32 Bottling was far from universally approved, however: Grigg states:

> Many retailers thought glass bottles a failure almost before they had really been tried: but often they were not properly washed and sterilised on return and the next lot of milk went sour. So before 1914, and indeed until 1930, in large towns like Bristol and Derby, a great deal of milk was delivered to customers in cans that were filled from a churn transported in a pony cart, of from a milk pail that was all shiny with brass bands outside, but opened at the top for use. The milk was dipped out with a long handled measure.
>
> (1989, p. 327)

33 J. Nugent Harris, 'The Organization of the Milk Supply', *Journal of the Board of Agriculture*, 1909–10, vol. XVI. See particularly Harris' discussion of the Wensleydale Pure Milk Society.

34 According to Nugent Harris:

> By organising the producing end of the business we eliminate risk of loss to the middlemen who are distributors in the big centres of consumption by giving them supplies to meet their requirements from day to day, instead of by the old system sending all the milk from the individual farms, whether the trader required it for his business or not.
>
> (ibid.)

35 The Co-operative Movement was one of the first to begin to discuss national control of milk supplies and of prices to producers, at its Congress in 1917. It continued to argue for national control throughout the 1920s. In 1929, the Co-operative Milk Trade Association was formed; it produced the plan for a National Milk Marketing Board which was approved by the National Farmers' Union and came into existence, via the Milk Marketing Act, three years later.

36 C. W. Walker-Tisdale, 'Milk Production', *Journal of the BDFA*, 1917, vol. XXXI, p. 30.

37 T. H. Middleton, for the Board of Agriculture, cited in Walker-Tisdale, 'Milk Production', pp. 30–1.

38 Walker-Tisdale, 'Milk Production', p. 29.

39 Primrose McConnell, 'Some Effects of the War on the Dairying Industry', *Journal of the BDFA*, 1917, vol. XXXI, p. 13.

40 Ibid., pp. 14–15.

41 Jenkins, *Drinka Pinta*, p. 74.

42 John Empson, 'The History of the Milk Marketing Board, 1933–1994: British Farmers' Greatest Commercial Enterprise', in J. Empson, *The UK Milk Marketing Boards: A Concise History*, Oxon: The Society of Dairy Technology, 2000, p. 9.

43 According to Empson:

> There were 141000 registered producers, including 70000 selling retail, with, on the other side of the coin, some 20000 buyers. All, perforce, had to be parties to the transfers and to the monthly transfers of cash. With these numbers, it was in fact a remarkable achievement that the Scheme got into effect at all.
>
> (2000, p. 13)

44 See Jenkins, *Drinka Pinta*, Jenkins' history was commissioned by and written on behalf of the National Milk Publicity Council to commemorate its 50th anniversary. Although it is therefore a very useful source empirically speaking for studying the history of milk publicity, because it represents an inside account of the development of milk advertising and the work of the NMPC, it is also a work of self-congratulation markedly lacking in critical or analytical depth. For this reason I base my account to some extent in this section upon a critical reading of Jenkins' history.

45 Lori Anne Loeb, *Consuming Angels: Advertising and Victorian Women*, Oxford: Oxford University Press, 1994, p. 8.

46 Francis T. Bond, 'Some Aspects of Milk and its Products from a Dietetic Point of View', *Journal of the BDFA*, 1893, vol. VIII, p. 33. See also Wilfred Buckley, *The Milk Supply: Report of a Conference Convened by the Health Committee of Farmers, Milk Dealers, and Others at the Town Hall in Bradford, May 4th 1915*, Manchester: W. M. Byles and Sons, 1915, p. 6.

47 H. Osborne, 'Health Week', in the *1925 Annual Report of the Medical Officer of Health for the City of Salford*, 1925, pp. 27–8.
48 H. Osborne, 'Public Health Propaganda', in the *1926 Annual Report of the Medical Officer of Health for the City of Salford*, 1926, pp. 79–90.
49 H. Osborne, 'Health Propaganda', in the *1928 Annual Report of the Medical Officer of Health for the City of Salford*, 1928, pp. 7–11.
50 Jenkins, *Drinka Pinta*, pp. 93–4.
51 George A. Auden, 'An Experiment into the Nutritive Value of an Extra Milk Ration', paper delivered at the Royal Sanitary Institute, Hull, 1923, stored at the Wellcome Library, Special Collections, WA1.
52 The Maternity and Child Welfare Act of 1918 made it a statutory responsibility for County and County Borough Councils to establish a Maternity and Child Welfare Committee and empowered them to provide ante-natal clinics, day nurseries, and free milk.
53 H. C. Corry Mann, *Diets for Boys of School Age*, Medical Research Council Report, London: HMSO, 1926.
54 H. C. Corry Mann, *A Comparison between the Diet of a Child in a Poor Industrial District and that of a Public School Boy in England*, Medical Research Council Report, London: HMSO, 1928.
55 The phrase 'propaganda for consumption' is taken from a paper presented to the World Dairy Congress in 1928 by J. G. Stapleton of the National Milk Publicity Council, under the title: 'Propaganda with regard to the Consumption of Milk'.
56 According to the August 1925 report of the Chairman of the National Milk Publicity Council.
57 Jenkins, *Drinka Pinta*, p. 101.
58 Ibid., p. 32.
59 Ibid., p. 104.
60 Ibid., p. 33.
61 See, for example, Gerald Leighton and Peter McKinlay, *Milk Consumption and the Growth of Schoolchildren: A Report on an Investigation in Lanarkshire Schools*, Department of Health for Scotland: HMSO, 1930. Also see J. Boyd Orr, *Milk Consumption and the Growth of Schoolchildren: Preliminary Report on Tests to the Scottish Board of Health*, London: HMSO, reprinted from *The Lancet*, 28 Jan. 1928.
62 Boyd Orr, J., *Food, Health and Income: Report on a Survey of Adequacy of Diet in Relation to Income*, London: Macmillan, 1936.
63 Empson, *The UK Milk Marketing Boards*, p. 13.
64 Ibid., p. 105.
65 As early as 1915 the Chairman of the National Clean Milk Society was stating that:

> The very first thing that strikes me about milk in this country is that it isn't used enough. I am saying something nobody can contradict when I tell you that the majority of people in this country do not recognise the full value of milk as food . . . We want more milk used in this country.
>
> (Buckley, 1915, p. 6)

66 This can be seen particularly clearly in the case of liquid milk, because it involves special difficulties of sanitary transportation and distribution: 'The most important disadvantage we farmers work under is that we are handling a commodity which it is most difficult to keep clean, and it is generally impossible to see the actual contamination of milk' (Buckley, *The Milk Supply*, p. 7).
67 Key to the development of this approach to economy-as-culture has been the work of Michel Callon. See Michel Callon, Yuval Millo and Fabian Muniesa (eds) *Market Devices*, Oxford: Blackwell, 2007.

3 Culture, order and disease in late nineteenth-century British dairying

1 Latour states:

> The smallest AIDS virus takes you from sex to the unconscious, then to Africa, tissue cultures, DNA and San Francisco, but the analysts, thinkers, journalists and decision makers will slice the delicate network traced by the virus for you into tidy compartments where you will find only science, only economy, only social phenomena, only local news, only sentiment, only sex.
>
> (1993, p. 2)

2 See, for example, Susan Sontag, *Illness as Metaphor*, London: Vintage, 1978. For a useful overview of medical anthropology, see Cecil Helman, *Culture, Health and Illness*, Oxford: Butterworth-Heinemann, 2000.

3 For a powerful account of the theological nature of secularism, see John Milbank, *Theology and Social Theory: Beyond Secular Reason*, Oxford: Blackwell, 1990. Milbank argues that the historical roots of what has become the secular sphere are the product of distinctly theological discourse, and that the conditions of possibility of the secular are in fact profoundly religious.

4 Michel Foucault, *Discipline and Punish: The Birth of the Prison*, London: Penguin, 1991, pp. 29–30. Foucault refers to the corporeal governing of 'souls', but humans must first be constituted as creatures *with* souls, in a dialectic which simultaneously denies souls to nonhumans.

5 For an account of Thomas Aquinas' anthropocentrism, see Andrew Linzey, *Animal Theology*, London: SCM, 1994, pp. 13–14. See also Richard Ryder, *Animal Revolution: Changing Attitudes towards Speciesism*, Oxford: Berg, 2000, p. 35.

6 This is typically overlooked by the ecology paradigm, which tends to be rooted in anthropocentric and homogeneous categories of 'nature' and 'society', despite its emphasis upon their inter-relations.

7 The tendency to privilege inherited concepts over material being and to acknowledge the latter only insofar as it may be compatible with the presumed coherence of the former is always indefensible, but particularly from a genealogical point of view.

8 For a critical account of the 'Great Chain of Being', see Peter Singer, 'All Animals Are Equal', in Tom Regan and Peter Singer (eds) *Animal Rights and Human Obligations*, Englewood Cliffs, NJ: Prentice Hall, 1976, p. 159.

9 All ontological discourse is tendentially theological in its form, for it involves the separation of being and practice into distinct and transcendental domains in an *a priori* division of what is only *a posteriori* divisible. Hence it fails to recognize the processes of attribution which give rise to what it posits as given.

10 Duckham states:

> The origin of these diseases not being traced, they were considered to have arisen spontaneously, and persons who entered upon the very precarious occupation of farming were told that they must be ready to submit to such diseases as a part of the ordinary casualties of their business.
>
> (1893, p. 266)

11 According to Duckham:

> Valuable as agricultural statistics are now generally considered to be, the collection of them was at the outset by no means popular, and the first Returns of 1866 are most incomplete. Since that date they have been year by

> year rendered more reliable, and through them the enormous value of the interest at stake has been ascertained.
>
> (ibid., p. 277)

12 The development of agricultural statistics then, through giving birth to the concept of animal 'population', simultaneously gave rise to what Latour calls 'the indisputable conflict between health and wealth', albeit the health of nonhumans. See Bruno Latour, *The Pasteurization of France*, Cambridge, MA: Harvard University Press, 1988, p. 16.

13 Brown states:

> There is no difficulty in realising the fact that an animal suffering from infectious disease is capable of communicating the disease by the agency of the tissues and fluids of the body.
>
> (1887, p. 47)

14 Duckham, 'Contagious Diseases of Livestock', p. 281.

15 Emphasizing this attitudinal shift, G. T. Brown remarked:

> Had the Privy Council of five-and-forty years ago attempted to arrest the progress of pleuro-pneumonia and foot-and-mouth disease by slaughter and stopping the free movement of animals, they would have raised a storm of opposition which would have rendered the effort futile.
>
> (1893, p. 283)

16 According to Duckham:

> The losses which have been sustained through the ravages of imported diseases among the flocks and herds of the United Kingdom since the first outbreak of that very infectious and contagious disorder popularly termed 'foot and mouth disease' until the present day cannot be estimated, inasmuch as no complete records were obtained until after the Contagious Diseases (Animals) Act of 1878 came into operation. That they amount in the aggregate to hundreds of millions sterling is beyond question. Ruinous as these losses were to stock owners, and seriously as the cost of meat to the consumers was enhanced by the repeated and severe outbreaks of the disease. No attempt was made by the Legislature to deal with the latter until after the visitation of cattle plague in 1865, and then only by half hearted measures, the outbreaks being represented to be spontaneous and the tracing of their origin not even troubled about.
>
> (1893, p. 262)

17 One contemporary agricultural reformer took it upon himself to compile estimates for his own county of Herefordshire, sending printed forms into every parish and hamlet with a circular letter asking for the information. He points out himself that the returns were incomplete, and suggests that this 'probably arose from the fear that advantage might be taken of those who had not made the proper returns in the first instance' (Ibid., p. 272).

18 McFadyean states:

> A tuberculous animal is a sort of hotbed for the propagation of the tubercle bacillus, and wherever such an animal exists it is a source of danger to other animals by reason of the bacilli which may be voided from its body.
>
> (McFadyean, 1895, p. 657)

19 Brown states:

> The importation of sheep affected with *sheep-pox* in 1847 created great alarm among flockmasters on account of the rapid spreading and fatal character of the disease, and in 1848 a Bill was introduced which subsequently became law. This Act, 11 and 12 Victoria, cap. cv. September 1848, gave powers to the Queen in Council to prohibit the importation of sheep, cattle, and other animals, or to admit them after quarantine, for the prevention of contagious diseases. On the same date the Act 11 and 12 Victoria, cap. cvii., came into operation, and provided for the passing of Orders in Council to prohibit or regulate the movement of animals, including horses; also of meat, hides, fodder, and other things likely to spread contagion. It further made it an offence to expose diseased sheep in markets . . . But a candid critic reading the Act would be forced to admit that its provisions were directed specially to the prevention of sheep-pox, which is the only disease referred to in the Act. It is true that general powers referring to other animals were conferred on the Privy Council. The powers were, however, limited to the prohibition of movement of animals, and of substances which might carry infection.
>
> (Brown, 1893, p. 282)

20 Brown states:

> Cattle-Plague was introduced in 1865, and began to spread extensively owing to the absence of any restrictions on the movement of diseased or affected animals. Some time was spent in making inquiries as to the nature of the disease, and an Order was passed calling upon owners of diseased cattle to give notice of the existence of disease; this, however, they neglected to do, preferring to send their stick to market as soon as a case of cattle plague was detected.
>
> (ibid., p. 283)

21 In reference to the proceedings of a Commons Select Committee on cattle diseases in 1873, to which he himself gave evidence, T. Duckham reported that 'Evidence of much weight, both by agriculturalists and professional witnesses, was given, tending to show that such enactments as would be required to exterminate foot-and-mouth disease would meet such strong opposition that they would be difficult, if not impossible, to carry out.' The Committee therefore recommended 'that the Privy Council should cease to issue orders for the check of this disease' (Duckham, 1893, p. 273).

22 In 1878, the Contagious Diseases (Animals) Act was passed into law. The Act did not include tuberculosis among the contagious diseases of animals. It would be some time more before tuberculosis received legislative attention.

23 In reference to Georges Cuvier's curatorship of France's National Museum of Natural History in the nineteenth century, John Pickstone explains that, for Cuvier,

> to understand how animals were alive was to understand the relations of structural and functional relations between organs, *and*, additionally and crucially, the functional interactions between each animal system and its *milieu*. This is the origin of our meaning of 'environment' as more than just 'surroundings'.
>
> (2000, p. 116)

24 McFadyean states:

> It hardly need be pointed out that a tuberculous animal will be a hundredfold

> more dangerous to its neighbours when it is housed with them than when it is grazed with them in the open air, or that overcrowding and imperfect ventilation will greatly favour the chance of the bacilli voided from a diseased animal finding its way into the body of another . . . It has recently been shown that direct sunlight is very speedily fatal to the germs of tuberculosis, and hence, even when bacilli are voided out of doors, they are probably very soon rendered innocuous. On the other hand, the darker a byre is the more favourable it is for the spread of tuberculosis.
>
> (1895, p. 658)

25 The Dairies and Cowsheds Orders of 1885 and 1899 required a certain minimum standard of accommodation, with specifications covering air space, light, ventilation and drainage. These orders were not universally acted upon by local authorities, however, and were widely held to be at best partially adequate:

> That the regulations in force, which are mainly those made under the Dairies, Cowsheds and Milkshops Orders of 1885 and 1899, are not universally deemed to be sufficient may be inferred from the fact that nearly one hundred local authorities, mostly borough councils, have by private legislation obtained further powers for dealing with the milk produced or coming into their respective areas . . . Up to a year ago, some 327 out of 1,794 borough or district councils had made no regulations under the Order of 1885, and consequently it may not unreasonably be assumed that practically little supervision or control is exercised over the milk supply of these districts.
>
> (Middleton, 1909, p. 51)

26 This points towards the fact that disciplinary power observes no ontological distinction between human and nonhuman in what Foucault refers to as its 'art of distributions', but rather produces this distinction through its own techniques, which arise in the practice of various disciplinary institutions before becoming generalized; such techniques must themselves then be seen as generative of the principles of ontological differentiation, of subject and object, culture and nature, human and nonhuman. See Michel Foucault, *Discipline and Punish: The Birth of the Prison*, London: Penguin, 1991, pp. 142–3.

27 This is what is referred to as 'the structure of the absent referent', in Carol J. Adams, *The Sexual Politics of Meat: A Feminist-Vegetarian Critical Theory*, New York: Continuum, 1996, pp. 40, 42, 44, 47.

28 'By the end of the nineteenth century dairy farming was becoming a more and more important branch of English farming . . . The number of milch cows kept in the country was perhaps larger than it had ever been' (Fussell, 1966, p. 56).

29 Even in 1890, one agricultural writer was able to say that 'The extent to which tuberculosis prevails among cattle in this country is unknown, as no attempt has hitherto been made to compile reliable statistics' (Duguid, 1890, p. 311). Similarly, six years later, it was still possible to complain that

> We have no reliable statistics by which to estimate the prevalence of tubercular udder disease in our cow stock, but we are convinced that thousands of quarts of milk daily become contaminated with the virus of tubercle from this source, and are retailed to the public, or converted into one or another of the various milk products.
>
> (Wortley Axe, 1896, p. 51)

30 So pervasive were these technologies that one dairy farmer, writing in the *Journal of the British Dairy Farmers' Association*, felt moved to complain:

> One might almost conclude that milk has had enough attention paid to it by 'experts' as from the cow to the consumer (and before and after) it is watched with solicitude; it is filtered, cooled, heated, bottled, sterilised, humanised, municipalised, and otherwise so maltreated that I am sure no self-respecting cow would own it; and an ordinary dairy farmer may be forgiven for enquiring into the necessity for all this.
>
> (Marshall, 1906, p. 17)

31 The first attempt to bottle milk was made by George Barham of the Express Dairy in 1884. The bottle was sealed with a wired-on cap which did not work in practice.

32 The turn towards breeding based on milk-yield recording and the systematic use of approved bulls marked a significant departure from the long-standing practice of selecting cattle by their outward appearance, which had even been made into a quasi-scientific principle in the form of Guernon's 'escutcheon' theory, according to which the milk-producing capacity of a cow could be determined by carefully examining the direction of growth of the hairs on the udder and the part of the cow between the thighs and hips. See Fussell, *The English Dairy Farmer*, pp. 29–30.

33 'The report of the Royal Commission appeared in November, but, whilst it brought no relief, it confirmed the worst opinions that had been expressed by Professor Simmonds and others that the pole-axe was the only remedy.' Indeed 'the whole of the veterinary profession were held up to ridicule for their ignorance in not being able to do more than prescribe the pole-axe and advocate the option of rigid precautionary measures.' Duckham, 'Contagious Diseases of Livestock', pp. 268–9.

34 The Adulteration of Food Acts of the 1860s provided some minimal protection against abuses such as watering-down by town retailers of milk. The Public Health Act of 1875 made compulsory the inspection of milk exposed for sale in England and Wales; the Sale of Food and Drugs Act of the same year forbade the addition to milk of anything 'injurious to health'. It was some time, however, before the means were in place to enforce these Acts in anything like an effective or consistent manner.

35 Brown states:

> It has recently been clearly proved that the milk of animals suffering from tuberculosis contains the tubercle bacilli in abundance, and it is an undoubted fact that such milk given to calves produces a tuberculous deposit in the lungs, and in other parts of the system.
>
> (Brown, 1887, p. 47)

All mention of the risk to humans is conspicuously absent. By 1896, however, the Veterinary Surgeon to the British Dairy Farmers' Association was confident in saying that:

> Tuberculosis is a deadly disease to which a large number of the human population succumb. Experiment has shown it to be transmissible from the cow to the lower animals by the agency of specifically infected milk, and both observation and theoretical consideration warrant the conclusion that it is so communicated to the human family.
>
> (Wortley Axe, 1896, p. 51)

Controversy continued, however, even among 'experts', and as late as 1909 the Consulting Chemist to the BDFA was applauded for declaring that 'He was one of the few men in England who believed that tuberculosis in cattle had nothing to do with the disease of the same name in human beings' (Minutes of the 1909 Midlands Dairy Conference, *Journal of the BDFA*, 1909, vol. XXIII, p. 34).

36 'These representations have always been made in the interest of public health, based on the idea that tuberculous meat and milk were dangerous if consumed as food' (Duguid, 1890, p. 316).

37 Thus culture/nature dualism also underpinned the bifurcation of society and economy, as economic processes became increasingly abstracted from social life with the further development of capital, becoming a 'second nature' of seemingly nonhuman forces. Hence the modern capitalist 'economy' was articulated in terms derived from the humanist notion of a 'nature' separate from 'culture'.

38 'Until some solution is arrived at in connection with the question of tuberculous meat and milk being used as food, there seems little probability that anything will be done as regards the disease of the living animal' (Duguid, 1890, p. 320).

4 Purifying milk: knowledge, sanitation and discipline

1 John Law and Annemarie Mol, 'The Actor Enacted: Cumbrian Sheep in 2001', in Lambros Malafouris and Carl Knappett (eds) *Material Agency: Towards a Non-Anthropocentric Approach*, New York: Springer, 2007.

2 Timothy Mitchell, *Rule of Experts: Egypt, Techno-Politics, Modernity*, Berkeley, CA: University of California Press, 2002, p. 21.

3 Patrick Joyce, *The Rule of Freedom: Liberalism and the Modern City*, London: Verso, 2003, p. 155; Simon Gunn, *The Public Culture of the Victorian Middle Class: Ritual and Authority in the English Industrial City 1840–1915*, Manchester: Manchester University Press, 2000.

4 Manchester was a very large city at the end of the nineteenth century with around 900,000 milk consumers. A century previously it had depended upon supplies brought along the Bridgewater Canal by cart or produced within the city. Its consumption in 1897 was about 14.7 million gallons per annum, only about 16 gallons a head or under 2.46 pints a week. The supply was mainly composed of milk brought by rail from Cheshire, Derby, Stafford, Lancashire and Salop, and delivered to stations in and within a 4-mile radius of the city amounting to 10.36 million gallons annually. Cows kept in the same area produced 2.8 million gallons, and a further supply of 1.5 million was brought in by cart from a radius of about 8 miles.

5 For a fuller exposition of this approach to method, see John Law, *After Method: Mess in Social Science Research*, London: Routledge, 2004.

6 'A great deal has been written in the press about milk being disastrous to infantile life in large centres.' Statement by Professor Nuttall 1909, cited in 'Report on the Midlands Dairy Conference', *Journal of the British Dairy Farmers' Association*, 1909, vol. XXIII, p. 73.

7 Savage states:

> It is unnecessary to offer proof of the connection, as the evidence is abundant, convincing, and accessible. At least 10 percent of average mixed milk samples contain tubercle bacilli, while probably about 15 to 20 percent of the tuberculosis in children is due to milk infection.
>
> (William G. Savage, 'The Protection of the Milk Supply', 1917.)

8 See the comments of Professor Nuttall after James Sadler, 'How to Secure a Milk Supply Free from Tuberculosis – from a Practical Farmer's Point of View', *Journal of the BDFA*, 1909, vol. XXIII, p. 73. See also J. T. Horner, 'The Milk Supply: From the Distributors' Point of View', *Journal of the BDFA*, 1909, vol. XXIII,

p. 36. Finally see the sixth resolution of the BDFA in the 'Report on the Midlands Dairy Conference', p. 84.

9 See R. Henry Rew, 'Contemporary Agricultural Law', *Journal of the Royal Agricultural Society of England*, 1914.

10 See J. P. Sheldon, 'Milk Selling Farmers and Coming Legislation', *Journal of the BDFA*, 1909, vol. XXIII, pp. 29–30.

11 Sheldon states:

> The avowed object of such legislation as we are led to expect is, primarily, the protection of infant life in towns against tuberculosis, which is alleged to come from the country. But have the officials concerned in it solved the problem which underlies the infection to which these town bred children are so unfortunately liable? Have they analysed the air of the great towns, and especially of the slums therein contained? Were they to devote to this potent source of contamination one half of the scrutiny they have given to milk, it is more than probable that their eyes would be opened to a different phase of the question . . . this infectious air will, when adequately investigated, be found to be more in fault than any supply of milk from the country.
>
> (ibid.)

12 'In this country Woodhead, McFadyean and I have for over twenty years collected evidence and conducted investigations indicating that bovine tuberculosis is very prevalent in Great Britain and is a danger to human health.' Sheridan Delépine, 'Tuberculosis and the Milk Supply', 1911.

13 Norman Hay Forbes, *Tuberculosis in Cattle – Its Relation to our Meat and Milk Supply: A Lecture to Agriculturalists*, Tunbridge Wells: R. Clements, 1899, p. 3.

14 The attitude of sanitary officials to this predicament was typically unsympathetic: 'To eradicate tuberculosis is expensive, but on economic grounds and merely from the point of view of the agricultural community, it pays to spend money for this purpose' (Savage, 1917, p. 8).

15 Keir Waddington, 'The Science of Cows: Tuberculosis and the State in the United Kingdom 1890–1914', *History of Science*, 2001, vol. 39, 3 (125), p. 355.

16 The Manchester Milk Clauses of 1899 were in fact a few years behind Glasgow's adoption of similar measures, reflecting Scotland's consistent position at the forefront of major reforms of the milk industry, such as milk recording. I have focussed upon Manchester on the grounds that although it was not chronologically the first city in Britain to introduce systematic milk testing linked to farm inspection, it developed by far the most influential model of this system. Glasgow was an advanced but isolated development, whereas Manchester, just a few years later, provided the catalyst for the adoption of similar measures in every city: 'Glasgow was about the first local authority to acquire powers to inspect dairies and cattle outside its own area from which milk was sent to that city. Manchester quickly followed, and it was over the Manchester Bill that the chief battle was fought.' C. B. Davies, 'Legislation Affecting the British Farmer', *Journal of the BDFA*, 1906, vol. XX, p. 86.

17 I should emphasize that I am not positing a one-man view of historical transformation here; neither do I want to subscribe to some variant of a structuralist account. My discussion of Delépine is intended to side-step such theoretical dichotomies as far as possible. Quite simply, the documents suggest that he was a key figure in the operationalization of milk testing and farm inspection in Manchester, and that its effectiveness in Manchester was pivotal in its subsequent application nationally. I leave the question of how truly 'determining' his influence was to one side. What matters for my purposes is what took place in the field of events, for which Delépine is merely a convenient narrative hook, not dissimilar to

the role of Pasteur in Bruno Latour's (1988) account of *The Pasteurization of France*.

18 Delépine, 'Tuberculosis and the Milk Supply', p. 22.

19 Ibid., p. 30.

20 According to Delépine:

> The Borough authorities have for a long time been able to control the farms and dairies within its boundaries, but up to 1899 were unable to influence adequately the rural districts from which the bulk of the milk supply was derived. Our preliminary investigations, conducted before that date, have shown that the rural supply was highly infected.
>
> (Delépine, 1911, p. 13)

21 '*Manchester Method* (Niven 1899) – The method of control at present in use in Manchester was devised for the purpose of protecting the town against the importation of tuberculous milk from the country' (ibid.).

22 Ibid., p. 14.

23 The discovery of tuberculin as a diagnostic agent was a late but significant development, providing the first reliable means to identify tuberculous animals. It was still not in use by the time of the Paris Congress on Tuberculosis in 1888.

24 Michel Foucault, *Discipline and Punish*, p. 197.

25 Bruno Latour, *We Have Never Been Modern*, p. 11.

26 Sheridan Delépine, *Pathogenic Properties of and Amount of Dirt Contained in the Milk Supplied to the Town of Manchester during the Ten Years 1897 to 1906 Inclusive*, Manchester: Sheratt and Hughes, 1908, p. 3.

27 Foucault, *Discipline and Punish*, p. 16.

28 Delépine, 'Tuberculosis and the Milk Supply', p. 21.

29 Ibid., p. 26.

30 'From a business point of view it is clear that if the funds available yearly are insufficient for effecting permanent good all over the country it is better to spend the money available each year in doing one part of the work thoroughly than to waste most of it in imperfect and discouraging general attempts' (ibid., p. 30).

31 'This work, to be effective, must not be left at the mercy of *voluntary notifiers*, and should be carried out under the supervision of competent veterinary surgeons entrusted with the duty of inspecting systematically all the cattle in the district under their charge' (ibid., p. 29).

32 'What became evident from the late nineteenth century were conceptions of the city as a *whole*, a new sense of the city as a totality that often embraced it in terms of its own larger, spatial "environment" ' (Joyce, 2003, p. 171).

33 Delépine often speaks explicitly of 'the utilization of scientific facts for administrative purposes' and 'results that have proved useful for administrative purposes'. See Delépine, 1911, pp. 15, 24.

34 Hay Forbes states:

> Let us for a moment enquire as to the direct cause of tuberculosis, and when we have determined that, we will be in a better position to take note of any accessory causes. In 1883 Dr. Robert Koch discovered that tuberculosis was caused by the presence of a very minute germ or fungus, which under the microscope has the appearance of a 'little rod' with rounded ends and is striped on its surface. This, when it attacks the tissues of the body, develops small round nodules or 'tubercles'. These germs or tubercle bacilli will only grow under certain conditions.
>
> (Hay Forbes, 1899, p. 4)

35 Delépine, 'Tuberculosis and the Milk Supply', 1911, p. 8.

36 'Obligatory passage point' is an ANT concept intended to capture the way in which certain elements in a network/assemblage become so durable and entrenched that they are able to circumscribe the possibilities of further network assemblage; they become 'hard' nodes or indispensable configurations in the network which 'narrow' the channels through which other agents may pass. This explains the apparent anteriority of certain 'realities'.

37 According to Delépine:

> It has been suggested that *resistance to tuberculous infection could be so increased* by improved social conditions (including good feeding, proper housing, and good sanitation) as to render the chances of infection slight. It is quite true that poor, weak, ill-fed individuals, crowded in unhealthy dwellings, are much more liable to suffer from infection than others; but I do not believe that simple sanitary and hygienic measures – valuable adjuncts though they are – can overcome the danger of infection by massive doses of infecting material, and, more especially, that infants could be protected in this way from the effects of feeding on tuberculous milk.
>
> (1911, p. 13)

38 See Joyce, *The Rule of Freedom*, pp. 63, 171.

39 Chris Otter, 'The Vital City: Public Analysis, Dairies and Slaughterhouses in Nineteenth-Century Britain', *Cultural Geographies*, 2006, vol. 13, pp. 517–37.

40 'Much confidence has been placed in microscopical examination . . . But when the bacilli are few the labour needed for their detection is so considerable as to render the thing impracticable' (Delépine, 1911, p. 11).

41 'As a matter of fact, tubercular milk is often remarkably free from bacteria capable of growing rapidly on gelatine or agar media. This kind of bacterial standard is absolutely useless for the purpose of preventing the sale of tuberculous milk' (ibid., p. 12).

42 Delépine states:

> The importance attached by many observers and administrators to *the number of bacteria present in milk* may be justified if the cultivation method is only used for the purpose of ascertaining how the milk has been handled, how long and at what temperature it has been kept. But unless the milk is tested by cultivation immediately after it has left the cow, this method does not give any material information as to the state of the cow.
>
> (ibid., p. 11)

43 Delépine, *Pathogenic Properties of and Amount of Dirt Contained in the Milk supplied to the Town of Manchester during the Ten Years 1897 to 1906 Inclusive*, 1908, p. 3.

44 Delépine states:

> When proper care is exercised thus method yields very reliable results, its great defect is that it cannot yield these results in less than ten days, and that on average about three weeks are necessary to ensure, for practical purposes, the reliability of negative results.
>
> (Delépine, 1911, p. 12)

45 In Delépine's view:

> The methods of testing which I have described are, I believe, the only ones that have been used extensively up to the present day. I think I am justified

> in saying that they are all unsatisfactory when employed for the purpose of controlling directly the sale of milk. I will go further and say that even if an almost instantaneous test were discovered by which the presence of tubercle bacilli in milk could be infallibly detected, this method would not solve the difficulty in a satisfactory manner. In the first place, the infectious material would continue to be distributed as long as the examination of a sample had not revealed its existence. Secondly, the only effect of the testing would be either the exclusion of the contaminated supply from a certain market and its diversion to another, or if all the markets were closed to tuberculous products the farmer, being left unguided, would, in the present state of things, be exposed to such losses that the dairy industry would suffer greatly. This method of control of tuberculous milk would therefore afford only partial protection to the consumer, and would seriously damage the dairy industry unless the farmers were at the same time helped to put their herds in order. This difficulty may, however, be partly overcome by a combination of the testing of milk with the inspection of farms. I will refer to this plan later on when giving an account of the work done in Manchester.
>
> (ibid., p. 12)

46 Delépine notes:

> The magnitude of the work is clearly shown by the statistics which have been collected during the past thirty years, and which prove the great prevalence and wide distribution of bovine tuberculosis. It is also clearly indicated by the results which we have obtained in Manchester in the course of the last fourteen years by the application of a method of control which may be considered as forming a link between municipal and agricultural methods.
>
> (ibid., p. 13)

47 Savage notes:

> Bacteriological standards as a basis upon which to accept or reject milk are undesirable, but they are invaluable as a *basis for classifying the milk producers* and for *judging the cleanliness of production*. Those producers whose milk contains abundance of bacteria require to be promptly visited, their methods of milking inquired into and their cowshed conditions inspected. In this way the producers of the dirtiest milk can be rapidly ascertained, and if legal powers are available the conditions altered and brought up to the standard of others.
>
> (1917, my italics)

48 'The legislation that permits one authority to enter upon a dairy premises situated in the area of another and sometimes distant authority caused a great outcry, and a very determined stand was made against it' (Davies, 1906, p. 87).

49 Horner, 'The Milk Supply: from the Distributor's Point of View', p. 36.

50 Report on The Midlands Dairy Conference, *Journal of the BDFA*, 1909, vol. XXIII, pp. 34, 82.

51 Ibid., p. 74.

52 Ibid., p. 76.

53 Sidney Villar, 'The Prevention of Tuberculosis in Dairy Cattle', *Journal of the BDFA*, 1909, vol. XXIII, p. 78.

54 This ontological and epistemological conflict was also expressed in farmers' frequent ridicule of what they viewed as the scientific experts' apparent lack of

practical knowledge: 'Unfortunately, our educational system seems "so careful of the type" of half-fledged scientists that for one who succumbs to the logic of facts a dozen start into being, each one with a more advanced – shall I say wilder? – theory than his predecessor' (Marshall, 1906, p. 17).

55 Delépine, 'Tuberculosis and the Milk Supply', p. 3.

56 Ibid.

57 'Although Koch's statement caused serious difficulties for a time, they led a great number of observers in almost every country to undertake careful investigations, which have considerably strengthened the views which had been challenged by him' (ibid., p. 5).

58 William Power, Stanley Griffith, Louis Cobbett, and Arthur Eastwood, *Final Report of the Royal Commission Appointed to Inquire into the Relations of Human and Animal Tuberculosis*, London: HMSO, 1911, p. 37.

59 Ibid., p. 36. Delépine regarded this as a vindication of his approach:

> That the outcome of ten years' work and controversy should be the entire confirmation of the views which a few of us have steadily defended in this country for more than twenty years is all the more gratifying in that it affords full justification for the action which has been taken by the authorities who have trusted our teaching.
>
> (Delépine, 1911, p. 8)

and

> We can accept as proved that *bovine tuberculosis is a serious source of danger to man, and that of all bovine tuberculous products tuberculous milk is the most dangerous* owing to the share it takes in the feeding of infants.
>
> (ibid.)

5 Milk and modernity Part II: measurement, rationalization and control

1 That rationalization involves a reconfiguration of human–nonhuman relations is implied by George Ritzer, perhaps the best-known contemporary sociologist of rationalization, when he says that it 'involves an increase in efficiency, predictability, calculability and control *through the substitution of non-human for human technology*' (Ritzer, *The McDonaldization Thesis: Explorations and Extensions*, London: Sage, 1998, Preface, my italics). For Ritzer, this amounts to dehumanization, but it is only so from the point of view of a residual Weberian humanism, for technology is very human. In fact, rationalization is just as much a process of humanist purification as dehumanization. In the field of human–animal relations, for example, this substitution of technological mediation for more direct forms of human–animal interaction means solidifying the separation of human and nonhuman and purifying each as a discrete domain, thereby engendering the notional autonomy of human 'culture'.

2 James Macintosh, 'Milk Recording', *Journal of the British Dairy Farmers' Association*, 1929, vol. XLI, p. 64.

3 Ibid., p. 65.

4 Primrose McConnell, 'Lessons from my Milk Register'; see Robert Shanks, 'Lessons from Milk Recording', *Journal of the BDFA*, 1930, vol. XLII, p. 68.

5 Macintosh, 'Milk Recording', p. 65.

6 Macintosh states:

> The practice of keeping records of the milk yield of individual cows was adopted by a small number of progressive dairy farmers towards the middle

> of the last century. The British Dairy Farmers' Association and several Agricultural Colleges and County Councils did excellent pioneer work in this direction, but the movement was confined to a few centres until the initiation of a National Milk Recording Scheme by the Ministry of Agriculture in 1914.
>
> (1921, p. 38)

See also James Macintosh, 'The Judging of Dairy Cows', *Journal of the BDFA*, 1938, vol. L, p. 35.

7 William Lawson, 'Milk Records', *Journal of the BDFA*, 1913, vol. XXVII, pp. 58–9.
8 Ibid., p. 60.
9 Shanks notes:

> The trophies go to the men who are at the height of production, understand their job, and want no support or encouragement to develop their herd. Milk recording is out to encourage the men with small and unprofitable yields, and guide them along lines which are to their advantage.
>
> (Shanks, 1926, p. 11)

10 Homogenization and autonomization are not inherently contradictory movements; the two may be part of a single process in which the disciplinary operation of the political centre is incrementally dispersed throughout the territory and engineered into 'non-political' forms, such as material structures, technologies, and scientific ways of knowing. Hence the political centre becomes present everywhere, a suitable metaphor for which would be the liberal rule of law, as distinct from simple sovereignty. It increasingly operates invisibly through the non-political, enabling its explicit, visible operation from the centre to gradually withdraw or move on to the establishment of new domains of control. This is the nature of modern governmentality, in which the advance of narrowly political 'freedom' is measured only by its material, epistemological, and social hollowing out; it is itself contained within a 'political' system which is both wider and deeper, invested in a multiplicity of 'expertise' and technical details, and which not only transcends the domain of freedom but is its genealogical condition of possibility.
11 Macintosh, 'Views and Reviews: Milk Recording', p. 38.
12 Macintosh, 'Milk Recording', p. 68.
13 Ibid., p. 69.
14 Ibid.
15 Macintosh notes:

> The argument may be advanced that the knowledge and experience possessed by dairy farmers in the older dairying districts make recording unnecessary, and that they can manage their herds economically and profitably without milk records. A moderate degree of success can, no doubt, be attained, but a sufficient number of farmers of long experience have taken up recording and found it a helpful and profitable innovation to make this point of view untenable. The degree of success attained without the aid of milk recording, can be intensified by making use of the information which milk records supply in the detection and elimination of unprofitable cows, in more economical feeding, and as a guide to improved methods of breeding.
>
> (1929, pp. 70–1)

16 Macintosh, 'Views and Reviews: Milk Recording', p. 41.
17 Macintosh, 'Milk Recording', p. 71.

18 Ibid., p. 72.
19 Alan Jenkins, *Drinka Pinta: The Story of Milk and the Industry that Serves it*, London: Heinemann, 1970, p. 39.
20 Sadler states:

> Milk, unlike cheese, is not a manufactured article (except in the sense that the cow is the manufacturer), and is not to any appreciable extent under the control of the producer, and no farmer, however clever he may be, can ensure that the milk will be up to a given standard. The most he can do is to guarantee that when it leaves his hands it is genuine.
>
> (1907, p. 22)

21 Sadler again notes:

> There may be said to be three common forms of adulteration: a) Abstraction of cream; b) Addition of water; c) Addition of separated milk. Now so long as one or all of these forms of adulteration are practised judiciously and strictly within limits, it is well-nigh impossible to 'prove' that an act of adulteration had actually taken place.
>
> (ibid., p. 24)

22 Ibid.
23 'Here is a natural product which has been – rightly or wrongly – standardized. The cow which gives it is nothing if not capricious, and varies the quality of her product without so much as a by your leave' (Ibid., p. 19). See also Sadler's comments in the *Journal of the BDFA*, 1909, vol. XXIII, 34. Finally, see J. T. Horner, 'The Milk Supply: From the Distributor's Point of View', *Journal of the BDFA*, 1909, vol. XXIII, p. 39.
24 For more on trials of animals, see E. P. Evans, *The Criminal Prosecution and Capital Punishment of Animals*, London: William Heinemann, 1906. For a contemporary take on this, see Jan Bondeson, *The Cat Orchestra and the Elephant Butler*, Stroud: Tempus, 2006, pp. 141–65.
25 Macintosh, 'Milk Recording', p. 66.
26 Shanks, 'Lessons from Milk Recording', p. 72.
27 Macintosh, 'Milk Recording', p. 66.
28 'Generally speaking, practical wisdom dictated that the proper treatment of a cow was simply to give suitable food and water at regular times, allow sufficient exercise for health, and keep the animal clean and warm' (Fussell, 1966, p. 53).
29

> A vast amount of information is placed before every farmer who keeps a milk record, if he knows how to use it. From it he may derive much information as to the influence of the different food on milk production, and by carefully taking note when a change of either quantity or kind of food is given what are the influences on the milk yield, may be saved a large amount of money.
>
> (Lawson, 1913, p. 63)

30 According to Fussell.

> Towards the end of the century supplements were being fed even in such favoured counties as Cornwall, where a large proportion of cereal food was used in winter, e.g. maize meal, crushed oats, barley meal and crushed wheat, although the cows were out in the fields all day and often all night.
>
> (1966, p. 53)

31 Fussell again:

> From around this time [1843] onwards the chemistry of animal nutrition was studied by Lawes and Glibert, Voelcker, and many others with close attention and increasing exactness. All kinds of feeding stuffs were analysed and valued according to their composition, and many suggestions were made for the rationing of livestock including the increasingly important dairy cow. The results obtained were published in such periodicals as the *Journal of the Bath and West of England Society*, the *Journal of the Royal Agricultural Society of England* [JRASE] and reprinted in popular papers like the *Farmers' Magazine* and many others. No doubt they were discussed by many smaller and more local clubs and societies of farmers. The material found its way into textbooks of both the recondite and popular sorts. From all these sources farmers could learn of the progress made in the novel science of rationing their animals with the food that would do them the most good, and so make them profitable.
>
> (Ibid., p. 49)

32 Ibid., p. 52.

33 Primrose McConnell states:

> The enquiries as to the feeding of cows which has been evolved as a sort of side issue to the keeping of milk records has opened up a whole kingdom of knowledge, and in nothing more than in regard to the cost of keeping a cow. Many farmers who have been feeding cows all their lives and who had arrived at what they considered a satisfactory method or methods of rationing their animals have been astonished when the records worked out showed them that the cost was greater than the average of many farms or than that of their neighbours, and have been induced to set their house in order accordingly.
>
> (1917, p. 88)

34 Fussell states:

> What appears certain is that a great many dairy farmers managed their cattle in very much the same way after 1850 as they had done before, though their feeding was helped by the increasing use of oil cake, linseed and imported maize.
>
> (1966, p. 50)

35 'No member of a milk recording society, nor, indeed, any dairy farmer throughout the length and breadth of the country, need be in ignorance as to whether he is feeding wisely and economically' (Macintosh, 'Milk Recording', p. 67).

36 'One notes, of course, that the question of disease crops up with unhappy frequency at all times when the breeding of dairy cattle is being discussed' (Newsham, 1913, vol. XXVII, p. 14).

37 Fussell states:

> Johnson quoted the Rev. W. L. Rham, who in turn had, I imagine, his data from Albrecht Thaer, that a cow required 3% of her weight per diem in food to keep her in good health. He adopted Thaer's standard of 100lb. of good hay, a somewhat insubstantial and variable basis for calculating the values of other foods. For example, clover hay in blossom equalled 90 when the hay was 100; mangold 339; linseed cake 69; rye and barley chaff, 179; and so on. This was the dawn of the scientific feeding of livestock.
>
> (1966, p. 49)

38 Shanks notes:

> Of all breeds the Friesian is undoubtedly the animal which can be most influenced in the conversion of food into milk. An instance is given in Germany, where by forced feeding a herd's output was increased by 65 per cent . . . Such cases as these are experimental, and have no bearing on what really happens on the average farm. Amidst it all we must pity the poor cow.
>
> (1930, p. 71)

39 Fussell states:

> In both Somerset and Cheshire it had been found that yields were not being kept up, presumably because of the continual abstraction of phosphates from the pastures as milk. In Cheshire some attempt to combat this was made by using bone manure. Somerset farmers were advised in 1851 as they had been in 1797 to combine arable and dairy, growing roots and artificial grasses to feed their cows. They shared a fairly general opinion that three acres of pasture were necessary to keep a cow, i.e. 1.5 acres of good land for the summer, and 2 tons of hay each for the winter as well as aftermath grazing.
>
> (1966, p. 48)

40 Macintosh, 'Milk Recording', pp. 67–8. For a similar point about milk quality and recording, see Macintosh, 'Views and Reviews: Milk Recording', p. 39.

41 Shanks notes:

> The clerical work with the record sheets entails irksome work for most farmers. The best scheme, from this point of view, is that adopted in Scotland where the farmer has no figuring; the whole of this is done by the specially trained recorders. At the end of the year, the total weight of milk and fat percentages are handed to the herd owner, the productive value of each animal being expressed by the total fat she produces during her milking period. This work, which has been going on for nearly a quarter of a century, was initiated by the efforts of the late John Speir.
>
> (1930, p. 69)

42 Fussell, *The English Dairy Farmer*, p. 18.

43 Ibid., p. 30.

44 'Each writer, though directing attention especially to utility points, could hardly avoid reference to other characteristics usually found in the breed or type of which he had most experience' (Mackintosh, 1938, p. 31).

45 Fussell, *The English Dairy Farmer*, p. 27.

46 Ibid., p. 29.

47 Ibid., p. 31.

48 Shanks notes:

> Breeding for milk is a lottery, even where the greatest care of selection takes place. We have but to refer to herds where this has been going on for the last 25 years, and note how slow progress has been. We hear of the remarkable results which are obtained in some herds, but this is due to sheer good luck as much as anything else.
>
> (1930, pp. 74–5)

49 Macintosh, 'Views and Reviews: Milk Recording', p. 40.

50 He states:

> It has been shown by a study of herd records that a good milking cow is not *necessarily* a breeder of good dairy stock. The ability of a cow to *transmit* her productive qualities is distinct from the *possession* of those qualities. She may pass them on or she may not. It is by milk records that we find out the cows which are good in themselves, and also to what extent their progeny inherit their good qualities.
>
> (ibid., p. 40)

51 Fussell, *The English Dairy Farmer*, pp. 29–30.
52 Ibid., p. 30.
53 Newsham states:

> The exaggerated importance of 'show' points was limited as soon as our Association [the BDFA] introduced milking and butter trials at the London Dairy Show. That these trials materially assisted in the discomfiture of those who upheld inspection prizes there is little reason to doubt, although one must admit that fashion still plays a great part among breeders of dairy stock.
>
> (1913, p. 14)

54 Macintosh notes:

> The results of milking trials and butter tests at shows were, however, of only limited value and by far the most important step in providing the information needed to check the accuracy of opinions on the value of dairy characteristics was the introduction of systematic milk recording.
>
> (1938, p. 35)

55 For a historical account of the breeding of pedigree animals as symbols of wealth and status, see Harriet Ritvo, *The Animal Estate: The English and Other Creatures in the Victorian Age*, Cambridge, MA: Harvard University Press, 1987.
56 Macintosh, 'Views and Reviews: Milk Recording', pp. 38–9.
57 Macintosh notes:

> One of the causes contributing to the rapid growth of recording was the high prices realised by cows with high certified records. The period of phenomenally high prices for non-pedigree cows with high records is past, but good recorded cows have undoubtedly shown less depreciation than most other live stock.
>
> (ibid., p. 41)

58 Ibid.
59 'The principal object of the scheme is to authoritatively indicate the most profitable and least profitable cows in each herd, so that the owners might discard the latter and breed only from the former' (Lawson, 1913, p. 61). 'The great object of milk recording is, of course, to find out which are the best yielding cows and which are the bad ones that should be eliminated' (McConnell, 1917, p. 87).
60 It is also true that animals are increasingly subjected to some of the same processes as citizens of modern capitalist societies, or rather, as workers in modern capitalist economies. This follows from the fact that a key part of the armoury of disciplinary techniques which make up modern governmentality is a micro-technology of control of the body, and the species of the bodies concerned is essentially irrelevant at this level; what matters is their usefulness to the capitalist economy.

However, rather than having humans and animals as givens in my analysis and looking at how they are treated by modern society, I have wanted to focus upon how relations between animal species are transformed into relations between 'humans' and 'animals', and between 'culture' and 'nature', and how these groups are constituted through their definition. Hence, the processes of modernization which involve treating animals like citizens in some respects, such as their abstract economic individualization through milk recording, involve at the same time the materialization of the very categories which insist upon an absolute distinction between humans and nonhumans. Hence, for example, animals could be treated in a manner partly akin to humans in the economic sphere in proportion to the degree to which the relations in this sphere were themselves naturalized, for this allowed the sociality of natural relations, which is to say the simultaneity of species relations and social relations, to continue to go unrecognized. Such individuation in the sphere of politics or of justice, for example was conspicuously absent, and over time showed an inverse relation to such developments within the economy.

61 Macintosh, 'Views and Reviews: Milk Recording', p. 41.

62 See Michel Foucault on 'anatomo-politics' and 'bio-politics' in Paul Rabinow (ed.) *The Foucault Reader*, London: Penguin, 1991, p. 262.

63 Economic theory aspires most closely to this model in the 'social sciences', although it is highly questionable whether this is a reflection of its object rather than an intrinsic part of the reification which constitutes that object.

6 Beyond 'culture' and 'nature': towards a post-humanist knowledge

1 See Ferdinand de Saussure, *Course in General Linguistics*, Waukegan, IL: Fontana, 1978.

2 My critique of semiotic logic is informed by three main influences: Vološinov's brilliant but still hugely under-acknowledged *Marxism and the Philosophy of Language*, Cambridge, MA: Harvard University Press, 1973; Fredric Jameson's similarly under-read *The Prison-House of Language*, Princeton, NJ: Princeton University Press, 1975; and Ludwig Wittgenstein's later philosophy of language.

3 John Law, 'Actor Network Theory and Material Semiotics', 2007, available at http://www.heterogeneities.net/publications/Law-ANTandMaterialSemiotics.pdf.

4 On this point I have in mind in particular Mitchell's paper, 'Future Measures: How to Calculate – for a World without Oil', presented at the CRESC Annual Lecture, University of Manchester, 14 May 2009. This showed how the particular material properties of oil, including the extreme technical difficulties of measuring or even accurately estimating the quantities of oil reserves 'below ground', have been key enabling factors in the emergence of an abstracted neo-liberal economics in which questions of value are conceptually divorced from material questions of resource use and finitude.

5 See also Webb Keane, 'Signs Are Not the Garb of Meaning: On the Social Analysis of Material Things', in Daniel Miller (ed.) *Materiality*, Durham, NC: Duke University Press, 2005.

6 Even acknowledging that human designs have entered into the corporeality of some animals through breeding, this distinction stands. Breeding may share some of the characteristics of technological design, but it is not the same thing. Taking a cue from Tim Newton, it may even be that the difference is best understood temporally, with breeding necessarily operating over a far longer temporal duration. See Tim Newton, *Nature and Sociology*, London: Routledge, 2007.

7 A very similar distinction is made by Kevin Rowlands, 'A Materialist Approach to Materiality', in Miller (ed.) *Materiality*.

8 Tim Newton, *Nature and Sociology*, London: Routledge, 2007.

9 Ludwig Wittgenstein, *Philosophical Investigations*, Oxford: Blackwell, 2001, paragraph 190.

10 Ibid., paragraphs 7, 10, 190.

11 See Vanessa Lemm, *Nietzche's Animal Philosophy: Culture, Politics, and the Animality of the Human Being*, New York: Fordham University Press, 2009.

12 The concept derives from Theodor Adorno, *Negative Dialectics*, London: Routledge, 1973. It describes a dialectic in which there is no positive stage of Hegelian 'synthesis' between opposing principles, but only a thinking of these principles 'at the same time', even in their contradiction. Hence a negative dialectic does not purport to 'transcend' opposition, but rather, more modestly, to translate the very logic of the opposition into a new and more complex whole, which still retains the irreconcilable principles as such. Though this is widely taken to be a critical re-reading of dialectics, in many ways it is simply a correct reading of Hegel's *Phenomenology of Spirit*.

13 Donna Haraway, *Simians, Cyborgs and Women: The Reinvention of Nature*, London: Free Association Books, 1991, p. 11.

14 This criticism does not extend to the broader interdisciplinary field of 'animal studies'.

15 I would exempt from this tendency recent work in cultural geography, which has tended to be better able to acknowledge the embeddedness of society within nature, and which under the rubric of 'hybrid geographies' has recently developed innovative ways to think about society-nature interfusion. See, in particular, Sarah Whatmore, *Hybrid Geographies: Natures, Cultures, Spaces*, London: Sage, 2002; Steve Hinchliffe, *Geographies of Nature: Societies, Environments, Ecologies*, London: Sage, 2007.

16 There is a very sizeable and rapidly growing literature, which I cannot possibly do justice to here. A few starting points for interested readers might include: William McGrew, *The Cultured Chimpanzee: Reflections on Cultural Primatology*, Cambridge: Cambridge University Press, 2004; Christophe Boesch, 'Complex Co-operation Among Tai Chimpanzees', in Frans De Waal and Peter Tyack (eds) *Animal Social Complexity: Intelligence, Culture, and Individualized Societies*, Cambridge, MA: Harvard University Press, 2003; Sarah Mesnick, Karen Evans, Barbara Taylor, John Hyde, Sergio Escorza-Trevino, and Andrew Dizon, 'Sperm Whale Social Structure: Why it Takes A Village to Raise a Child', in De Waal and Tyack (eds) *Animal Social Complexity*; Jan Aram Van Hoof and Signe Preuschoft, 'Laughter and Smiling: The Intertwining of Nature and Culture', in De Waal and Tyack (eds) *Animal Social Complexity*.

17 See, for example, Toshisada Nishida, 'Individuality and Flexibility of Cultural Behaviour Patterns in Chimpanzees', in De Waal and Tyack (eds) *Animal Social Complexity*; Susan Perry, 'Coalitionary Aggression in White-Faced Capuchins', in De Waal and Tyack (eds) *Animal Social Complexity*; Gerald Wilkinson, 'Social and Vocal Complexity in Bats', in De Waal and Tyack (eds) *Animal Social Complexity*.

Bibliography

Primary sources

Aikman, C. M., 'Pasteurized Milk', *Journal of the British Dairy Farmers' Association*, 1896, vol. XI, Part I.

Auden, George A., *An Experiment into the Nutritive Value of an Extra Milk Ration*, paper delivered at the Royal Sanitary Institute, Hull, 1923, stored at the Wellcome Library, Special Collections, WA1.

Bond, Francis T., 'On Some Aspects of Milk and its Products from a Dietetic Point of View', *Journal of the British Dairy Farmers' Association*, 1893, vol. VIII.

Boyd Orr, J., *Milk Consumption and the Growth of Schoolchildren: Preliminary Report on Tests to the Scottish Board of Health*, London: HMSO, reprinted from *The Lancet*, 28 Jan. 1928.

—— *Food, Health and Income: Report on a Survey of Adequacy of Diet in Relation to Income*, London: Macmillan, 1936.

Brown, G. T., 'Milk in Relation to Infectious Diseases', *Journal of the British Dairy Farmers' Association*, 1887, vol. III, Part I.

—— 'The Progress of Legislation Against Contagious Diseases of Livestock: Part II', in *Journal of the Royal Agricultural Society of England*, 1893, Third Series, vol. IV.

Buckley, Wilfred, *The Milk Supply: A Report of a Conference Convened by the Health Committee of Farmers, Milk Dealers, and Others at the Town Hall in Bradford, May 4th 1915*, Manchester: W. M. Byles and Sons, 1915.

—— *The Milk Problem*, London: National Clean Milk Society, 1919.

Davies, C. B., 'Legislation Affecting the British Farmer', *Journal of the British Dairy Farmers' Association*, 1906, vol. XX.

Delépine, Sheridan, *Pathogenic Properties of and Amount of Dirt Contained in the Milk Supplied to the Town of Manchester during the Ten Years 1897 to 1906 Inclusive*, Manchester: Sheratt and Hughes, 1908.

—— 'Tuberculosis and the Milk Supply', paper presented to the Association of Veterinary Officers of Health Annual Meeting in Edinburgh, 13 October, 1911, published in *Journal of Meat and Milk Hygiene*, March 1912, stored in the John Rylands University of Manchester Library, Medical Pamphlets 2, 6: 34.

Duckham, T., 'The Progress of Legislation Against Contagious Diseases of Livestock: Part I', *Journal of the Royal Agricultural Society of England*, 1893, Third Series, vol. IV.

Duguid, W., 'Tuberculosis in Animals and its Relation to Consumption in Man', *Journal of the Royal Agricultural Society of England*, 1890, Third Series, vol. I.

Evans, E. P., *The Criminal Prosecution and Capital Punishment of Animals*, London: William Heinemann, 1906.

Harley, William, *The Harleian Dairy System*, London: J. Ridgeway, 1829.

Hay Forbes, Norman, *Tuberculosis in Cattle – Its Relation to our Meat and Milk Supply: A Lecture to Agriculturalists*, Tunbridge Wells: R. Clements, 1899.

Henry Rew, R., 'Contemporary Agricultural Law', *Journal of the Royal Agricultural Society of England*, 1914.

Hill, H., 'Liverpool – Last Stronghold of Town Cowkeepers', *Dairy Engineering*, April 1956.

Horner, J. T., 'The Milk Supply: From the distributors' point of view', *Journal of the British Dairy Farmers' Association*, 1909, vol. XXIII.

Lawson, William, 'Milk Records', *Journal of the British Dairy Farmers' Association*, 1913, vol. XXVII.

League of Nations, *Bulletin of the Health Organization*, 1937, vol. VI, no. 3, June.

Leighton, Gerald and Peter McKinlay, *Milk Consumption and the Growth of Schoolchildren: A Report on an Investigation in Lanarkshire Schools*, Department of Health for Scotland, London: HMSO, 1930.

Macewen, Hugh, *The Public Milk Supply*, London: Blackie, 1910.

Macintosh, James, 'Views and Reviews: Milk Recording', *Journal of the British Dairy Farmers' Association*, 1921, vol. XXXIII.

—— 'Milk Recording', *Journal of the British Dairy Farmers' Association*, 1929, vol. XLI.

—— 'The Judging of Dairy Cows', *Journal of the British Dairy Farmers' Association*, 1938, vol. L.

Mann, H. C. Corry, *Diets for Boys of School Age*, London: HMSO, 1926.

—— *A Comparison between the Diet of a Child in a Poor Industrial District and that of a Public School Boy in England*, Medical Research Council Report, London: HMSO, 1928.

McConnell, Primrose, 'The Best Form of Cow Stall', *Journal of the British Dairy Farmers' Association*, 1899, vol. XIV, Part II.

—— 'Some Effects of the War on the Dairying Industry', *Journal of the British Dairy Farmers' Association*, 1917a, vol. XXXI.

—— 'The Present Position and Future Developments of Dairy Education – Practice, *Journal of the British Dairy Farmers' Association*, 1917b, vol. XXXI.

—— 'Lessons from my Milk Register', cited in Robert Shanks, 'Lessons from Milk Recording', *Journal of the British Dairy Farmers' Association*, 1930, vol. XLII.

McFadyean, J., 'The Prevention of Tuberculosis in Cattle', *Journal of the Royal Agricultural Society of England*, 1895, Third Series, vol. VI.

Marshall, Charles, 'The Desirability or Otherwise of the Municipal Supply of Milk – From the Dairy Farmer's Point of View', *Journal of the British Dairy Farmers' Association*, 1906, vol. XX.

Middleton, Christopher, 'Suggested Amendments of the Dairies, Cowsheds, and Milkshops Orders in Relation to Buildings and Inspection', *Journal of the British Dairy Farmers' Association*, 1909, vol. XXIII.

Minutes of the 1909 Midlands Dairy Conference, Journal of the British Dairy Farmers' Association, 1909, vol. XXIII.

Moffitt, L. W., *England on the Eve of the Industrial Revolution*, London, 1925.

Newsham, J. C., 'The Breeding and Marketing of Dairy Cows', *Journal of the British Dairy Farmers' Association*, 1913, vol. XXVII.

Nugent Harris, J., 'The Organization of the Milk Supply', *Journal of the Board of Agriculture*, 1909–10, vol. XVI.

Osborne, H., 'Health Week', in the *Annual Report of the Medical Officer of Health for the City of Salford*, Salford, 1925.

—— 'Public Health Propaganda', in the *Annual Report of the Medical Officer of Health for the City of Salford*, Salford, 1926.

—— 'Health Propaganda', in the *Annual Report of the Medical Officer of Health for the City of Salford*, Salford, 1928.

Pocock, Sidney, 'Tuberculosis of Cattle – From the Dairyman's Point of View', *Journal of the British Dairy Farmers' Association*, 1898, vol. XIII, Part II.

Power, William, Stanley Griffith, Louis Cobbett, and Arthur Eastwood, *Final Report of the Royal Commission Appointed to Inquire into the Relations of Human and Animal Tuberculosis*, London: HMSO, 1911.

Sadler, James, 'Some Milk Problems', *Journal of the British Dairy Farmers' Association*, 1907, vol. XXI.

—— 'How to Secure a Milk Supply Free from Tuberculosis – from a Practical Farmer's Point of View', *Journal of the British Dairy Farmers' Association*, 1909, vol. XXIII.

Savage, William G., 'The Protection of the Milk Supply', lecture delivered at the Royal Institute of Public Health, 28 February 1917, published in the June 1917 *Journal of State Medicine*, stored as a paper in the John Rylands University of Manchester Library, Delépine collection.

Shanks, Robert, 'Where We Are and Where We Ought To Be', *Journal of the British Dairy Farmers' Association*, 1926, vol. XXXVIII.

—— 'Lessons from Milk Recording', *Journal of the British Dairy Farmers' Association*, 1930, vol. XLII.

Sheldon, J. P., 'Milk in Commerce', *Journal of the British Dairy Farmers' Association*, 1893, vol. III.

—— 'Milk Selling Farmers and Coming Legislation', *Journal of the British Dairy Farmers' Association*, 1909, vol. XXIII.

Speir, John, 'The Air of Cow Houses', *Journal of the British Dairy Farmers' Association*, 1896, vol. XI, Part I.

Stapleton, J. G., 'Propaganda with Regard to the Consumption of Milk', paper presented to the World Dairy Congress of 1928, on behalf of the National Milk Publicity Council.

Struthers, John, 'Milk and its Distribution', published in the *Journal of Meat and Milk Hygiene*, October 1911, stored as a paper in the John Rylands University of Manchester Library, Medical Pamphlets 1, Delépine 25:16.

Villar, Sidney, 'The Prevention of Tuberculosis in Dairy Cattle', *Journal of the British Dairy Farmers' Association*, 1909, vol. XXIII.

Walker-Tisdale, C. W., 'Milk Production', *Journal of the British Dairy Farmers' Association*, 1917, vol. XXXI.

Wortley Axe, J., 'Milk and Disease', *Journal of the British Dairy Farmers' Assocation*, 1892, vol. II.

—— 'Sources of Milk Infection', *Journal of the British Dairy Farmers' Assocation*, 1896, vol. XI, Part I.

Secondary sources

Abbott, A., *The System of Professions: An Essay on the Division of Expert Labour*, Chicago: University of Chicago Press, 1988.

—— *Chaos of Disciplines*, Chicago: University of Chicago Press, 2001.

Adams, Carol, *The Sexual Politics of Meat: A Feminist-Vegetarian Critical Theory*, New York: Continuum, 1996.

Adorno, Theodor, *Negative Dialectics*, London: Routledge, 1973.

Asdal, Kristin, 'On Politics and the Little Tools of Democracy: A Down to Earth Approach', *Distinktion: Scandinavian Journal of Social Theory*, 2008, No. 16, special issue on 'The Technologies of Politics', pp. 11–27.

Ashmore, Malcolm, 'Behaviour Modification of a Catflap: A Contribution to the Sociology of Things', *Kennis en Methode*, 1993, vol. 17, pp. 214–29.

Baker, Steve, *Picturing the Beast: Animals, Identity and Representation*, Chicago: University of Illinois Press, 2001.

Benton, Ted, 'Biology and Social Science: Why the Return of the Repressed Should Be given a Cautious Welcome', *Sociology*, 1991, vol. 25, no. 1, pp. 1–29.

Bhaskar, Roy, *The Possibility of Naturalism*, London: Routledge, 1998.

—— *A Realist Theory of Science*, London: Verso, 2009.

Bijker, W. and John Law (eds) *Shaping Technology, Building Society: Studies in Sociotechnical Change*, Cambridge, MA: MIT Press, 1992.

Boesch, Christophe, 'Complex Co-operation Among Tai Chimpanzees', in Frans de Waal and Peter Tyack (eds) *Animal Social Complexity: Intelligence, Culture, and Individualised Societies*, Cambridge, MA: Harvard University Press, 2003.

Bondeson, Jan, *The Cat Orchestra and the Elephant Butler*, Stroud: Tempus, 2006.

Braverman, Harry, *Labour and Monopoly Capital*, New York: Monthly Review Press, 1974.

Brooker, Peter, *A Concise Glossary of Cultural Theory*, London: Arnold, 1999.

Bryant, Clifton, 'The Zoological Connection: Animal-related Human Behaviour', *Social Forces*, 1979, 58(2), pp. 399–421.

Callon, Michel, 'Some Elements of a Sociology of Translation: Domestication of the Scallops and the Fisherman of St. Brieuc Bay', in John Law (ed.) *Power, Action and Belief: A New Sociology of Knowledge*, Sociological Review Monographs, London: Routledge and Kegan Paul, 1986.

Callon, Michel, with Yuval Millo and Fabian Muniesa (eds) *Market Devices*, Oxford: Blackwell, 2007.

Campbell, Ben, 'How Bio-technology Makes Human Kinship with Nature Visible', paper presented at *Animals and Science: Anthropological Approaches*, University of Manchester, 24 June 2005.

Carver, Terrell, 'Darwinism', in Tom Bottomore (ed.) *Dictionary of Marxist Thought*, Oxford: Blackwell, 1991.

Castree, Noel and Bruce Braun (eds), *Social Nature: Theory, Practice and Politics*, Oxford: Wiley-Blackwell, 2001.

Chambers Dictionary of Etymology, Edinburgh: Chambers, 1988.

Darwin, Charles, *The Origin of Species*, Hertfordshire: Wordsworth Editions, 1998.

Daston, Lorraine and Gregg Mitman (eds), *Thinking with Animals: New Perspectives in Anthropomorphism*, New York: Columbia University Press, 2006.

Debord, Guy, *The Society of the Spectacle*, London: Rebel Press/Dark Star, 1977.

Deleuze, Gilles and Félix Guattari, *A Thousand Plateaus: Capitalism and schizophrenia*, London: Continuum, 2004.

Delphy, Christine, *Close to Home: A Material Analysis of Women's Oppression*, London: Hutchinson, 1984.

Derrida, Jacques, *Spectres of Marx*, New York: Routledge, 1994.

—— *The Animal That Therefore I Am*, New York: Fordham University Press, 2008.

De Waal, Frans and Peter Tyack (eds) *Animal Social Complexity: Intelligence, Culture, and Individualised Societies*, Cambridge, MA: Harvard University Press, 2003.

Diamond, Jared, *Guns, Germs and Steel*, London: Vintage, 1998.

Dickens, Peter, *Society and Nature: Changing our Environment, Changing Ourselves*, Cambridge: Polity, 2004.

Durkheim, Emile, 'Sociology and its Scientific Field', in Emile Durkheim, *Essays on Sociology and Philosophy*, New York: Harper & Row, 1964.

Eagleton, Terry, *The Idea of Culture*, Oxford: Blackwell, 2000.

Empson, John, 'The History of the Milk Marketing Board, 1933–1994: British Farmers' Greatest Commercial Enterprise', in *The UK Milk Marketing Boards: A Concise History*, Oxon: The Society of Dairy Technology, 2000.

Fernandez-Armesto, Felipe, *So You Think You're Human?* Oxford: Oxford University Press, 2004.

Foucault, Michel, *The Order of Things: An Archaeology of the Human Sciences*, London: Tavistock, 1970.

—— *Discipline and Punish: The Birth of the Prison*, London: Penguin, 1991.

—— 'Truth and Power', in Paul Rabinow (ed.) *The Foucault Reader*, London: Penguin, 1991.

—— 'Governmentality', in Graham Burchell, Colin Gordon and Peter Miller (eds) *The Foucault Effect: Studies in Governmentality*, London: Harvester Wheatsheaf, 1991.

Fussell, G. E., *The English Dairy Farmer, 1500–1900*, London: Frank Cass & Co, 1966.

Gerth, H. H. and C. Wright Mills, 'Intellectual Orientations', in H. H. Gerth and C. Wright Mills (eds) *From Max Weber: Essays in Sociology*, London: Routledge, 1991.

Giddens, Anthony, *Capitalism and Modern Social Theory: An Analysis of the Writings of Marx, Durkheim and Weber*, Cambridge: Cambridge University Press, 1971.

—— *Durkheim*, Sussex: Harvester Press, 1978.

Grigg, David, *English Agriculture: An Historical Perspective*, Oxford: Blackwell, 1989.

Gunn, Simon, *The Public Culture of the Victorian Middle Class*, Manchester: Manchester University Press, 2000.

Haraway, Donna, *Simians, Cyborgs, and Women*, London: Free Association Books, 1991.

—— *Primate Visions: Gender, Race and Nature in the World of Modern Science*, London: Verso, 1992.

—— *The Companion Species Manifesto: Dogs, People, and Significant Otherness*, Chicago: University of Chicago Press, 2003.

Hegel, G. W. F., *Phenomenology of Spirit*, Oxford: Oxford University Press, 1977.

Helman, Cecil, *Culture, Health and Illness*, Oxford: Butterworth-Heinemann, 2000.

Hinchliffe, Steve, *Geographies of Nature: Societies, Environments, Ecologies*, London: Sage, 2007.

Jameson, Fredric, *The Prison-House of Language: A Critical Account of Structuralism and Russian Formalism*, Princeton, NJ: Princeton University Press, 1975.

Jenkins, Alan, *Drinka Pinta: The Story of Milk and the Industry that Serves It*, London: Heinemann, 1970.

Joyce, Patrick (ed.), *The Social in Question: New Bearings in History and the Social Sciences*, London: Routledge, 2002.

—— *The Rule of Freedom: Liberalism and the Modern City*, London: Verso, 2003.

Keane, Webb, 'Signs Are Not the Garb of Meaning: On the Social Analysis of Material Things', in Daniel Miller (ed.) *Materiality*, Durham, NC: Duke University Press, 2005.

Kilman, Andrew, 'Marx's Concept of Intrinsic Value', in *Historical Materialism*, 2000, No. 6.

Latour, Bruno, 'Give Me a Laboratory and I Will Raise the World', in Karin Knorr-Cetina and Michael Mulkay (eds) *Science Observed*, London: Sage, 1983.

—— *The Pasteurization of France*, Cambridge, MA: Harvard University Press, 1988.

—— *We Have Never Been Modern*, Harlow: Pearson, 1993.

—— *Reassembling the Social: An Introduction to Actor Network Theory*, Oxford: Oxford University Press, 2005.

—— *Politics of Nature: How to Bring the Sciences into Democracy*, Cambridge, MA: Harvard University Press, 2004.

Latour, Bruno and Steve Woolgar, *Laboratory Life: The Social Construction of Scientific Facts*, London: Sage, 1979.

Law, John, 'Introduction: Monsters, Machines and Sociotechnical Relations', in John Law (ed.) *A Sociology of Monsters? Essays on Power, Technology and Domination*, London: Routledge, 1991.

—— *Organizing Modernity*, Oxford: Blackwell, 1994.

—— *After Method: Mess in Social Science Research*, London: Routledge, 2004.

—— 'Actor Network Theory and Material Semiotics', 2007, available at http://hwww.heterogeneities.net/publications/Law-ANTandMaterialSemiotics.pdf

—— 'Culling, Catastrophe and Collectivity', *Distinktion: Scandinavian Journal of Social Theory*, 2008, No. 16, special issue on 'The Technologies of Politics', pp. 61–77.

—— and Annemarie Mol, 'The Actor Enacted: Cumbrian Sheep in 2001', in Lambros Malafouris and Carl Knappett (eds) *Material Agency: Towards a Non-Anthropocentric Approach*, New York: Springer, 2007.

Lee, Paula Young (ed.), *Meat, Modernity and the Rise of the Slaughterhouse*, Durham, NC: University of New Hampshire (Library Binding), 2008.

Lemm, Vanessa, *Nietzche's Animal Philosophy: Culture, Politics, and the Animality of the Human Being*, New York: Fordham University Press, 2009.

Linzey, Andrew, *Animal Theology*, London: SCM, 1994.

Loeb, Lori Anne, *Consuming Angels: Advertising and Victorian Women*, Oxford: Oxford University Press, 1994.

Lukács, Georg, *History and Class Consciousness*, London: Merlin, 1971.

Malafouris, Lambros and Carl Knappett (eds), *Material Agency: Towards a Non-Anthropocentric Approach*, New York: Springer, 2008.

Marx, Karl, *Capital: A Critique of Political Economy*, vol. I, London: Penguin, 1976.

—— 'Economic and Philosophical Manuscripts', in David McLellan (ed.) *Karl Marx: Selected Writings*, Oxford: Oxford University Press, 1977a.

—— 'Theses on Feuerbach', in David McLellan (ed.) *Karl Marx: Selected Writings*, Oxford: Oxford University Press, 1977b.

—— 'The German Ideology', in David McLellan (ed.) *Karl Marx: Selected Writings*, Oxford: Oxford University Press, 1977c.

—— 'Capital', in David McLellan (ed.) *Karl Marx: Selected Writings*, Oxford: Oxford University Press, 1977d.

—— 'A Contribution to the Critique of Political Economy', cited in David Andrews, 'Commodity Fetishism as a Form of Life', in Gavin Kitching and Nigel Pleasants (eds) *Marx and Wittgenstein: Knowledge, Morality and Politics*, London: Routledge, 2002.

McGrew, W. C., 'Ten Dispatches from the Chimpanzee Culture Wars', in Frans de Waal and Peter Tyack (eds) *Animal Social Complexity: Intelligence, Culture, and Individualised Societies*, Cambridge, MA: Harvard University Press, 2003.

—— *The Cultured Chimpanzee: Reflections on Cultural Primatology*, Cambridge: Cambridge University Press, 2004.

Mesnick, Sarah, and Karen Evans, Barbara Taylor, John Hyde, Sergio Escorza-Trevino, and Andrew Dizon, 'Sperm Whale Social Structure: Why it Takes A Village to Raise a Child', in Frans de Waal and P. Tyack (eds) *Animal Social Complexity: Intelligence, Culture, and Individualised Societies*, Cambridge, MA: Harvard University Press, 2003.

Michael, Mike, *Constructing Identities: The Social, the Nonhuman, and Change*, London: Sage, 1996.

—— *Reconnecting Culture, Technology and Nature: From Society to Heterogeneity*, London: Routledge, 2000.

Milbank, John, *Theology and Social Theory: Beyond Secular Reason*, Oxford: Blackwell, 1990.

Miller, Daniel, 'Materiality', in Daniel Miller (ed.) *Materiality*, Durham, NC: Duke University Press, 2005.

Mitchell, Timothy, *Colonising Egypt*, Berkeley, CA: University of California Press, 1992.

—— *Rule of Experts: Egypt, Techno-Politics, Modernity*, Berkeley, CA: University of California Press, 2002.

—— 'Future Measures: How to Calculate – for a World without Oil', paper presented at the *CRESC Annual Lecture*, University of Manchester, 14 May 2009.

Mol, Annemarie, 'Ontological Politics: A Word and Some Questions', in John Law and John Hassard (eds) *Actor Network and After*, Oxford and Keele: Blackwell and *Sociological Review*, 1999.

Moog, Sandra and Rob Stones (eds), *Nature, Social Relations and Human Needs*, Houndmills: Palgrave-Macmillan, 2008.

Newton, Tim, *Nature and Sociology*, London: Routledge, 2007.

Nimmo, Richie, 'Governing Nonhumans: Knowledge, Sanitation and Discipline in the Late 19th and Early 20th Century British Milk Trade', *Distinktion: Scandinavian Journal of Social Theory*, 2008, No. 16, special issue on 'The Technologies of Politics', pp. 77–99.

—— 'Auditing Nature, Enacting Culture: Rationalization as Disciplinary Purification in Early Twentieth Century British Dairy Farming', *Journal of Historical Sociology*, 2008, vol. 21, no. 2/3, pp. 272–302.

—— *An Historical Inquiry into the Social Dynamics of Animal Exploitation in Relation to Structures of Class and Gender*, Sociology Working Papers, University of Manchester, Department of Sociology, 2003.

Nishida, Toshisada, 'Individuality and Flexibility of Cultural Behaviour Patterns in Chimpanzees', in F. de Waal and P. Tyack (eds) *Animal Social Complexity:*

Intelligence, Culture, and Individualised Societies, Cambridge, MA: Harvard University Press, 2003.

Otter, Chris, 'The Vital City: Public Analysis, Dairies and Slaughterhouses in Nineteenth-Century Britain', *Cultural Geographies*, 2006, vol. 13, pp. 517–37.

Perry, Susan, 'Coalitionary Aggression in White-Faced Capuchins', in Frans de Waal and Peter Tyack (eds) *Animal Social Complexity: Intelligence, Culture, and Individualised Societies*, Cambridge, MA: Harvard University Press, 2003.

Pickering, Andrew, *The Mangle of Practice: Time, Agency and Science*, Chicago: University of Chicago Press, 1995.

Pickstone, John, *Ways of Knowing: A New History of Science, Technology and Medicine*, Manchester: Manchester University Press, 2000.

Pinney, Christopher, 'Things Happen: Or, From Which Moment Does That Object Come?' in Daniel Miller (ed.) *Materiality*, Durham, NC: Duke University Press, 2005.

Poovey, Mary, *Making a Social Body: British Cultural Formation 1830–1864*, Chicago: University of Chicago Press, 1995.

Rabinow, Paul (ed.), *The Foucault Reader*, London: Penguin, 1991.

Reuten, Geert, 'The Interconnection of Systematic Dialectics and Historical Materialism', *Historical Materialism*, 2000, no. 7, pp. 137–65.

Ritvo, Harriet, *The Animal Estate: The English and Other Creatures in the Victorian Age*, Cambridge, MA: Harvard University Press, 1987.

Ritzer, George, *The McDonaldization Thesis: Explorations and Extensions*, London: Sage, 1998.

Rose, Nikolas, *Powers of Freedom: Reframing Political Thought*, Cambridge: Cambridge University Press, 1999.

Rowlands, Michael, 'A Materialist Approach to Materiality', in Daniel Miller (ed.) *Materiality*, Durham, NC: Duke University Press, 2005.

Ryder, Richard, *Animal Revolution: Changing Attitudes towards Speciesism*, Oxford: Berg, 2000.

Saussure, Ferdinand de, *Course in General Linguistics*, Waukegan, IL: Fontana, 1978.

Schmidt, Alfred, *The Concept of Nature in Marx*, London: New Left Books, 1971.

Schutz, Alfred, *The Phenomenology of the Social World*, Evanston, IL: Northwestern University Press, 1967.

Shapin, Steven and Simon Schaffer, *Leviathan and the Air Pump: Hobbes, Boyle, and the Experimental Life*, Princeton, NJ: Princeton University Press, 1985.

Singer, Peter, 'All Animals Are Equal', in Tom Regan and Peter Singer (eds) *Animal Rights and Human Obligations*, Englewood Cliffs, NJ: Prentice Hall, 1976.

Smith, Tony, 'The Relevance of Systematic Dialectics to Marxian Thought', in *Historical Materialism: Research in Critical Marxist Theory*, 1999, No. 4, London: London School of Economics.

—— 'A Reply to Fine, Lapavitsas and Milonakis', in *Historical Materialism*, 2000, No. 6.

Sontag, Susan, *Illness as Metaphor*, London: Vintage, 1978.

Tucker, William, 'Max Weber's Verstehen', *Sociological Quarterly*, 1965, vol. 2, pp. 157–165, reprinted in Peter Hamilton (ed.) *Max Weber: Critical Assessments*, London: Routledge, 1991.

Van Hoof, Jan Aram and Signe Preuschoft, 'Laughter and Smiling: The Intertwining of Nature and Culture', in Frans de Waal and Peter Tyack (eds) *Animal*

Social Complexity: Intelligence, Culture, and Individualised Societies, Cambridge, MA: Harvard University Press, 2003.

Vološinov, V. N., *Marxism and the Philosophy of Language*, Cambridge, MA: Harvard University Press, 1973.

Waddington, Keir, 'The Science of Cows: Tuberculosis and the State in the United Kingdom 1890–1914', *History of Science*, 2001, vol. 39, no. 125, pp. 355–81.

Wahrman, Dror, *The Making of the Modern Self: Identity and Culture in Eighteenth-Century England*, New Haven, CT: Yale University Press, 2004.

Whatmore, Sarah, *Hybrid Geographies: Natures, Cultures, Spaces*, London: Sage, 2002.

Wheen, Francis, *Karl Marx*, London: Fourth Estate, 2000.

Wilkinson, Gerald, 'Social and Vocal Complexity in Bats', in Frans de Waal and Peter Tyack (eds) *Animal Social Complexity: Intelligence, Culture, and Individualised Societies*, Cambridge, MA: Harvard University Press, 2003.

Williams, Raymond, *Culture*, London: Collins, 1981.

Wittgenstein, Ludwig, *Philosophical Investigations*, Oxford: Blackwell, 2001.

Young, Robert, 'Nature', in Tom Bottomore (ed.) *Dictionary of Marxist Thought*, Oxford: Blackwell, 1991.

Index

9780415558747